Springer Theses

Recognizing Outstanding Ph.D. Research

For further volumes:
http://www.springer.com/series/8790

Aims and Scope

The series "Springer Theses" brings together a selection of the very best Ph.D. theses from around the world and across the physical sciences. Nominated and endorsed by two recognized specialists, each published volume has been selected for its scientific excellence and the high impact of its contents for the pertinent field of research. For greater accessibility to non-specialists, the published versions include an extended introduction, as well as a foreword by the student's supervisor explaining the special relevance of the work for the field. As a whole, the series will provide a valuable resource both for newcomers to the research fields described, and for other scientists seeking detailed background information on special questions. Finally, it provides an accredited documentation of the valuable contributions made by today's younger generation of scientists.

Theses are accepted into the series by invited nomination only and must fulfill all of the following criteria

- They must be written in good English.
- The topic should fall within the confines of Chemistry, Physics, Earth Sciences, Engineering and related interdisciplinary fields such as Materials, Nanoscience, Chemical Engineering, Complex Systems and Biophysics.
- The work reported in the thesis must represent a significant scientific advance.
- If the thesis includes previously published material, permission to reproduce this must be gained from the respective copyright holder.
- They must have been examined and passed during the 12 months prior to nomination.
- Each thesis should include a foreword by the supervisor outlining the significance of its content.
- The theses should have a clearly defined structure including an introduction accessible to scientists not expert in that particular field.

Michael Prim

Polarization and CP Violation Measurements

Angular Analysis of B $\rightarrow$ ϕ K* Decays and Search for CP Violation at the Belle Experiment

Doctoral Thesis accepted by
Karlsruhe Institute of Technology, Germany

Springer

Author
Dr. Michael Prim
Institute for Experimental Nuclear Physics
Karlsruhe Institute of Technology
Karlsruhe
Germany

Supervisor
Prof. Michael Feindt
Institute for Experimental Nuclear Physics
Karlsruhe Institute of Technology
Karlsruhe
Germany

ISSN 2190-5053 ISSN 2190-5061 (electronic)
ISBN 978-3-319-35953-3 ISBN 978-3-319-05756-9 (eBook)
DOI 10.1007/978-3-319-05756-9
Springer Cham Heidelberg New York Dordrecht London

Printed on acid-free paper

Springer is part of Springer Science+Business Media (www.springer.com)

Supervisor's Foreword

The discovery of *CP* violation effects in B meson decays at B factories has led to the Nobel Prize in Physics in 2008 for its explanation in the Standard Model of particle physics. Meanwhile, precision measurements of such effects in special rare decays are a powerful way to search for New Physics effects beyond the Standard Model, complementary to the direct search at the LHC.

Michael Prim describes a thorough data analysis of the rare B meson decay into Phi K^* on data taken by the Belle Collaboration at the B-meson factory KEKB over 10 years. This reaction is interesting, because it in principle allows the observation of large *CP* violation effects. In the Standard Model, however, no *CP* violation is expected. An observation of *CP* asymmetries thus immediately implies new physics.

Michael Prim had to reconstruct the decays from four charged particles. For this complex analysis he developed and made publicly available several generally useful new analysis tools, particularly for quantifying multivariate dependence. He also developed a fast numerical evaluation procedure for acceptance integrals in the amplitude likelihood fit using some master integrals. Although not new, the particle physics community seems to have forgotten this technique over the last two decades. Exploiting this method, Michael Prim could accelerate systematic uncertainty calculations by several orders of magnitude. Also, this software he made publicly available.

From the 5-fold differential cross-section he altogether determined 26 parameters, all of them are either first or world-best measurements. Some of the final results are surprising. These findings now await theoretical explanation in low energy-QCD models. However, independently of these details, the observation of large strong phases proves a large sensitivity to new physics effects, but no signs of *CP* violation are observed. Thus, this work reinforces the validity of the Standard Model.

The thesis is well written and may serve as a textbook example of how to explore highest energies on the basis of low energy precision experiments and modern statistical data analysis exploiting quantum mechanical interference effects.

Karlsruhe, December 2013 Prof. Michael Feindt

Abstract

The Standard Model of particle physics describes the properties and fundamental interactions of matter. The theory has been tested by countless measurements and has proven to be a robust model. Yet, there exist hints on physics beyond the Standard Model. Deviations between experimental measurements and theoretical predictions or observations like the within the Standard Model unexplainable matter–antimatter asymmetry in the universe demand more precise measurements and searches.

The constituents of matter are leptons and quarks, the latter form bound states called hadrons such as the proton, neutron, or mesons. In this thesis, a measurement of the decay of a neutral B meson into a ϕ meson and an exited K^* meson ($B^0 \to \phi K^*$) is presented. In the Standard Model this transition is described via a single one-loop diagram, a higher order amplitude in the perturbation theory.

CP violation, an asymmetry in the behavior of matter and antimatter, is a result of interference effects and requires at least two contributing amplitudes. Therefore, no *CP* violation is expected within the Standard Model in $B^0 \to \phi K^*$ transitions. Any potential physics beyond the Standard Model could, however, contribute with a second amplitude to the transition that could result in a measurable deviation from the expectation. Measuring asymmetries that are sensitive to *CP* violation in the decay probes the Standard Model and indirectly constrains the possible parameter space for physics beyond the Standard Model.

In the $B^0 \to \phi K^*$ system, K^* denotes contributions from scalar (S-wave, spin $J = 0$), vector (P-wave, $J = 1$), and tensor (D-wave, $J = 2$) components from $(K\pi)^*_0$, $K^*(892)^0$, and $K^*_2(1430)^0$, respectively. As the ϕ is a vector meson, the pseudoscalar B^0 decays into a vector–scalar, vector–vector, or vector–tensor state. Conservation of angular momentum results in up to three polarizations (longitudinal, parallel, and perpendicular) for these decays, thus providing several quantities that are potentially sensitive to deviations from the Standard Model expectation. Using a partial wave analysis, which exploits the different angular distributions of the S-, P-, and D-wave component, the branching fraction $\mathcal{B}_J$, the longitudinal (perpendicular) polarization fraction f_{LJ} ($f_{\perp J}$), the relative phase of the parallel (perpendicular) amplitude $\phi_{\parallel J}$ ($\phi_{\perp J}$) to the longitudinal amplitude, the strong phase differences δ_{0J} between the S-wave and the P- and D-wave, and a number of parameters related to direct *CP* violation in all these quantities are measured. In total, 26 parameters are measured.

The analysis was performed using the full Belle data sample, consisting of an integrated luminosity of 711 fb^{-1} containing $(772 \pm 11) \times 10^6$ $\mathrm{B\overline{B}}$ pairs collected at the $\Upsilon(4S)$ resonance at the KEKB asymmetric-energy e^+e^- collider. The $\Upsilon(4S)$ resonance decays into a B meson pair and the subsequent decay products are detected and recorded by the Belle detector.

Several new and improved methods have been applied with respect to previous Belle measurements. Neural networks have been employed to obtain an observable that discriminates between signal and the dominating background from $e^+e^- \rightarrow q\bar{q}\,(q \in \{u, d, s, c\})$ events. A nine-dimensional maximum likelihood fit was used to perform the final parameter extraction. For this fit, a tool has been developed to obtain a reliable measure of dependence among the observables in multivariate data sets. Furthermore, a method that can improve the computation time of numeric integrations in partial wave analysis and amplitude analyses, in general, by orders of magnitude was developed.

The obtained results are summarized in Table 1 and supersede all previous Belle results for the P-wave component $B^0 \rightarrow \phi K^*(892)^0$. The analysis also provides the first measurement related to the S- and D-wave components $B^0 \rightarrow \phi(K\pi)^*_0$ and $B^0 \rightarrow \phi K^*_2(1430)^0$, respectively, at Belle. The results are consistent with other measurements from the BaBar collaboration and improve the uncertainties on all parameters related to the S- and P-wave components. Naive expectations predict a dominant longitudinal polarization in the decay, which is confirmed in $B^0 \rightarrow \phi K^*_2(1430)^0$ decays but in conflict with the result obtained in $B^0 \rightarrow \phi K^*(892)^0$ decays. All parameters related to *CP* violation, which are listed

Table 1 Summary of the 26 parameters measured in the $B^0 \rightarrow \phi K^*$ system

Parameter	$\phi(K\pi)^*_0$ $J=0$	$\phi K^*(892)^0$ $J=1$	$\phi K^*_2(1430)^0$ $J=2$
$\mathcal{B}_J\ (10^{-6})$	$4.3 \pm 0.4 \pm 0.4$	$10.4 \pm 0.5 \pm 0.6$	$5.5^{+0.9}_{-0.7} \pm 1.0$
f_{LJ}	...	$0.499 \pm 0.030 \pm 0.018$	$0.918^{+0.029}_{-0.060} \pm 0.012$
$f_{\perp J}$	...	$0.238 \pm 0.026 \pm 0.008$	$0.056^{+0.050}_{-0.035} \pm 0.009$
$\phi_{\parallel J}$ (rad)	...	$2.23 \pm 0.10 \pm 0.02$	$3.76 \pm 2.88 \pm 1.32$
$\phi_{\perp J}$ (rad)	...	$2.37 \pm 0.10 \pm 0.04$	$4.45^{+0.43}_{-0.38} \pm 0.13$
δ_{0J} (rad)	...	$2.91 \pm 0.10 \pm 0.08$	$3.53 \pm 0.11 \pm 0.19$
$\mathcal{A}_{CPJ}$	$0.093 \pm 0.094 \pm 0.017$	$-0.007 \pm 0.048 \pm 0.021$	$-0.155^{+0.152}_{-0.133} \pm 0.033$
$\mathcal{A}^0_{CPJ}$	...	$-0.030 \pm 0.061 \pm 0.007$	$-0.016^{+0.066}_{-0.051} \pm 0.008$
$\mathcal{A}^{\perp}_{CPJ}$	...	$-0.14 \pm 0.11 \pm 0.01$	$-0.01^{+0.85}_{-0.67} \pm 0.09$
$\Delta\phi_{\parallel J}$ (rad)	...	$-0.02 \pm 0.10 \pm 0.01$	$-0.02 \pm 1.08 \pm 1.01$
$\Delta\phi_{\perp J}$ (rad)	...	$0.05 \pm 0.10 \pm 0.02$	$-0.19 \pm 0.42 \pm 0.11$
$\Delta\delta_{0J}$ (rad)	...	$0.08 \pm 0.10 \pm 0.01$	$0.06 \pm 0.11 \pm 0.02$

The first error is statistical and the second due to systematic uncertainties in the bottom half of the table, are consistent with zero and the absence of *CP* violation.

The results of the measurement in $B^0 \rightarrow \phi K^*$ decays have been published in

M. Prim et al. (Belle Collaboration), "Angular analysis of $B^0 \rightarrow \phi K^*$ decays and search for *CP* violation at Belle," Physical Review D **88**, 072004 (2013).

The tool for measuring dependence in multivariate data sets has been published in

M. Feindt and M. Prim, "An algorithm for quantifying dependence in multivariate data sets," Nuclear Instruments and Methods in Physics Research A **698**, 84 (2013).

Acknowledgments

First of all I thank my doctoral advisor Prof. Michael Feindt for instructive discussions, advice, support, and the possibility to carry out this research in his working group.

I thank Prof. Dr. Günter Quast for being the second reviewer of this thesis.

I thank Dr. Thomas Kuhr and Dr. Anže Zupanc for their excellent supervision during my entire Ph.D.

I thank further Christian Pulvermacher, Bastian Kronenbitter, Dr. Benjamin Fuchs, and again Dr. Thomas Kuhr and Dr. Anže Zupanc for proofreading this thesis at different stages.

I thank the entire working group and the remaining Institute of Experimental Nuclear Physics for 4 years of excellent working atmosphere.

I thank the research training group "Hochenergiephysik und Teilchenastrophysik" of the German Research Foundation and the "Studienstiftung des Deutschen Volkes" for supporting and funding of my graduate studies and research.

Finally, I thank my parents Christine and Thomas Prim for their unconditional support in all circumstances.

Contents

Chapter 1
Introduction

The theory known as the Standard Model of particle physics describes the properties and fundamental interactions of matter. Certain experiments like the measurement of the anomalous magnetic moment of the muon, match the theoretical predictions with a precision of up to eleven digits [1]. Yet, a deviation has been observed with the available experimental and theoretical precision; being just one of many examples that indicate some physics beyond the Standard Model.

On the one hand, scientists are developing new models to explain deviations or unexpected observations. On the other hand, they are improving the experimental measurements to constrain these models. These days the ATLAS and CMS experiments at the Large Hadron Collider are, besides their search for the Higgs boson, pushing the energy frontier to a new level and aiming for direct observation of physics beyond the Standard Model. Other experiments pursue a complementary approach. Measurements like those of the anomalous magnetic moment of the muon are aiming for an indirect evidence of new physics by pushing experimental results to highest precisions.

The B-factory experiments Belle and BaBar have been and are still among the precision experiments in flavor physics, an important part of the Standard Model. B-factories provide a unique environment for measurements of B-mesons and the related electroweak properties of the Standard Model. Belle [2, 3] and BaBar [4, 5] measurements led to the confirmation of combined violation of charge-conjugation C and parity-transformation P in the B-meson system. The measurement of CP violation allows to distinguish between matter and antimatter. According to the CPT theorem, all Lorentz-invariant local quantum field theories are invariant under the combined transformation of CP-conjugation and time-reversal T. Therefore, CP violation allows to establish an arrow of time. The violation of CP symmetry is further assumed to be one of the necessary requirements to explain the observed matter-antimatter asymmetry in the universe [6]. Therefore the study of the smallest things may help to understand the biggest things.

This thesis presents an analysis of $\mathrm{B}^0 \rightarrow \phi \mathrm{K}^*$ decays with data collected at the Belle experiment, located at the Japanese High Energy Accelerator Research Organisation in Tsukuba, Japan. The properties of the decay are measured with an

M. Prim, *Polarization and CP Violation Measurements*, Springer Theses,
DOI: 10.1007/978-3-319-05756-9_1, © Springer International Publishing Switzerland 2014

angular analysis and in addition, direct *CP* violation in the decay is studied; both providing experimental results that are sensitive to the influence of possible physics beyond the Standard Model. The studied decay is a rare decay that is not expected to have a significant *CP*-violating asymmetry. However, it is sensitive to the possible influence of physics beyond the Standard Model that could result in sizable and measurable effects.

Chapter 2 briefly reviews the Standard Model of particle physics and gives a more detailed introduction into the theoretical concepts that are important for the study of $B^0 \to \phi K^*$ decays. Chapter 3 summarizes the experimental setup, consisting of the KEKB accelerator and the Belle detector. The methods that have been applied in the analysis are described in Chap. 4. The chapter includes existing methods as well as newly developed methods. The experimental reconstruction of the decay is described in Chap. 5, whereas the maximum likelihood fit model, the core component of the analysis, is discussed in Chap. 6. The results of the measurements are presented in Chap. 7. A summary and concluding remarks are given in Chap. 8.

1.1 Textbooks for Further Reading

The Chaps. 2 and 3 about theoretical and experimental basics have been kept brief with purpose when originally writing the thesis. The references given in the chapters for further reading provide often a more detailed but very compact set of information. As this publication of the thesis is aiming explicitly for students some additional textbooks references are provided below. These references should help students to get into the matter of this thesis.

Chapter 2

The author can recommend Refs. [7] and [8] for further reading, where both are suited for experimentalists but the first provides a more theoretical introduction in the field of particle physics. Two textbooks that go into the depth of quantum field theory can be found in Refs. [9] and [10].

Reference [11] is an excellent review about flavor physics at the Tevatron. It covers various topics such as mixing and *CP* violation from an experimentalist point of view but providing also the necessary theoretical background. Further, Ref. [12] is a great book about *CP* violation in general and not limited to the Bmeson system.

Chapter 3

A textbook about detector systems in general is given by Ref. [13]. The book has also a dedicated chapter about the Belle detector as an example of a general-purpose detector. Silicon detectors are covered in detail in Ref. [14].

References

1. G.W. Bennett et al., (Muon (g-2) Collaboration). Final report of the E821 muon anomalous magnetic moment measurement at BNL. Phys. Rev. D **73**(7), 072003 (2006)
2. K. Abe et al., (Belle Collaboration). Observation of large CP violation in the neutral B meson system. Phys. Rev. Lett. **87**, 091802 (2001)
3. I. Adachi et al., (Belle Collaboration). Precise measurement of the *CP* violation parameter $\sin 2\phi_1$ in $B^0 \to (c\bar{c})K^0$ Decays. Phys. Rev. Lett. **108**(17), 171802 (2012)
4. B. Aubert et al., (BaBar Collaboration). Observation of CP violation in the B^0 meson system. Phys. Rev. Lett. **87**, 091801 (2001)
5. B. Aubert et al., (BaBar Collaboration). Measurement of time-dependent *CP* asymmetry in $B^0 \to (c\bar{c})K^{(*)0}$ decays. Phys. Rev. D **79**(7), 072009 (2009)
6. A.D. Sakharov, Violation of *CP* invariance, *C* asymmetry and Baryon asymmetry of the universe. JETP **5**(1), 24–27 (1967)
7. D. Griffiths, *Introduction to Elementary Particles* (Wiley-VCH, Weinheim, 2008)
8. D. Perkins, *Introduction to High Energy Physics* (Cambridge University Press, Cambridge, 2000)
9. L. Ryder, *Quantum Field Theory* (Cambridge University Press, Cambridge, 1996)
10. M. Peskin, D. Schroeder, *An Introduction to Quantum Field Theory* (Perseus Books, Reading, 2006)
11. T. Kuhr, *Flavor Physics at the Tevatron: Decay, Mixing and CP-Violation Measurements in* $p\bar{p}$*-Collisions. Springer Tracts in Modern Physics,* vol. 249 (Springer, Berlin, 2013)
12. K. Kleinknecht, *Uncovering CP Violation: Experimental Clarification in the Neutral K Meson and B Meson Systems. Springer Tracts in Modern Physics,* vol. 195 (Springer, Berlin, 2004)
13. C. Grupen, B. Shwartz, *Particle Detectors. Cambridge Monographs on Particle Physics, Nuclear Physics, and Cosmology: 26* (Cambridge University Press, Cambridge, 2008)
14. F. Hartmann, *Evolution of Silicon Sensor Technology in Particle Physics. Springer Tracts in Modern Physics,* vol. 231 (Springer, Berlin, 2009)

Chapter 2
Theoretical Principles

In this chapter the theoretical principles of the Standard Model of particle physics are summarized. A brief introduction into the CKM matrix and direct *CP* violation will be given. Furthermore, the polarization of B meson decays and the idea of a partial wave analysis will be discussed in detail as it is the foundation for the analysis presented in this thesis.

2.1 The Standard Model of Particle Physics

The Standard Model (SM) of particle physics is the agreed common theory of particle physics. The core principles of the SM are dating back several decades, even as some predictions have not been measured and proven by experiments until recently. Over the decades, dozens of physicists have received the Nobel Prize for work related to the SM, not only theorists, but also experimentalists measuring its properties and inventing new instrumentation for these measurements.

The SM describes the properties and fundamental interactions of matter. The interactions are known as the strong, weak, and electromagnetic force. Gravitation, although being a fundamental force, is not part of the SM since it interacts on completely different scales as the other three. The SM is a quantum gauge theory with three local groups

$$SU(3)_C \otimes SU(2)_L \otimes U(1)_Y, \tag{2.1}$$

where C denotes the color charge of the strong interaction, L the left chirality of the weak interaction and Y the hypercharge of the electromagnetic interaction. The electromagnetic and weak interactions have been unified by the Glashow-Weinberg-Salam (GWS) model [1–3] in the 1960s to the electroweak theory, whereas the strong interaction has been first described in the 1970s by Gross, Wilczek, and Politzer [4–7] and is known as quantum chromodynamics (QCD).

M. Prim, *Polarization and CP Violation Measurements*, Springer Theses,
DOI: 10.1007/978-3-319-05756-9_2,

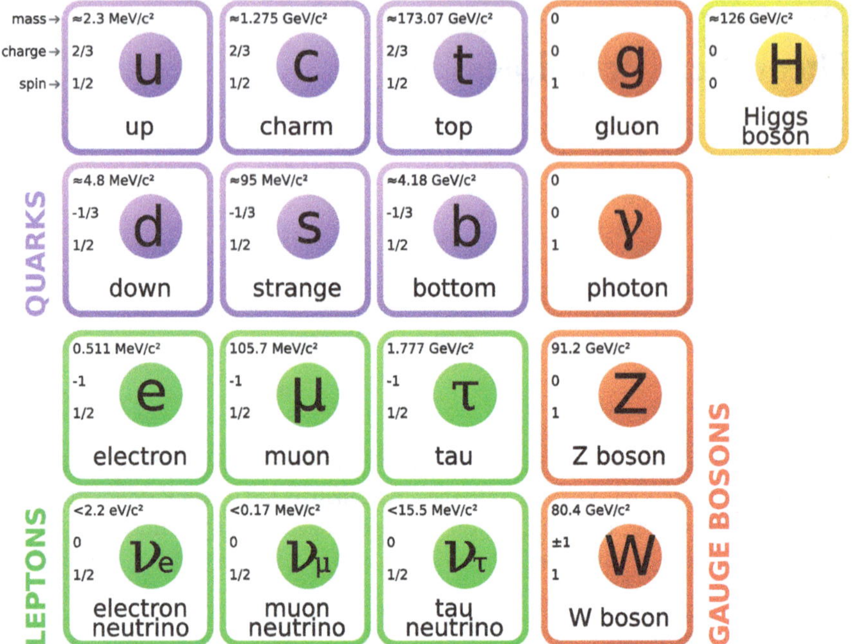

Fig. 2.1 Overview of the particles in the Standard Model of particle physics. Three generations (columns) of quarks and leptons are the constituents of matter, whereas the photon, gluons, and $W^{\pm}$ and Z^0 are the mediators of the electromagnetic, strong, and weak interaction, respectively. The Higgs boson is the gauge boson related to the Higgs mechanism that gives mass to the particles. Taken from Ref. [8]

All interactions have in common that the forces are mediated by bosonic exchange particles, the gauge bosons, which couple to the charge of the fermionic matter particles and, if charged, self-couple. The matter particles are divided into quarks and leptons, where quarks have color, weak, and electromagnetic charge and interact with all three forces. Leptons only have weak and electromagnetic charge, thus do not interact with the strong force. Neutrinos even lack the electromagnetic charge and their only interaction is by the weak interaction. The mediators of the strong, weak, and electromagnetic interaction are the gluons, $W^{\pm}$ and Z^0 bosons, and the photon, respectively. The $W^{\pm}$ and Z^0 bosons are also referred to as charged and neutral currents. An overview of the particles in the SM of particle physics and the gauge bosons is given in Fig. 2.1.

The GWS model describes the electroweak interactions as spontaneously broken symmetry groups

$$SU(2)_L \otimes U(1)_Y \rightarrow U(1)_{\mathrm{EM}}, \tag{2.2}$$

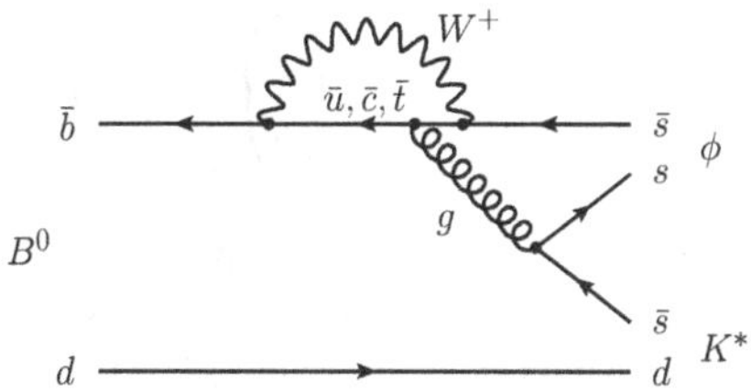

Fig. 2.2 Feynman diagram of the decay $B^0 \to \phi K^*$

which are broken by the Higgs mechanism [9] that generates the masses of the $W^\pm$ and Z^0 boson as well as the masses of the fermions. The Higgs boson related to this mechanism, has been discovered recently [10, 11] by the ATLAS and CMS experiments at the Large Hadron Collider. It completes the electroweak part of the SM, nearly thirty years after the discovery of the massive $W^\pm$ and Z^0 gauge bosons [12–14] by the UA1 and UA2 collaborations and about fifty years after the GWS model pointed out the direction.

All fermions have related anti-particles, with conjugated charges, e. g. the positron e^+ being right handed and having positive electrical charge. The quarks are not free particles, but as a consequence of QCD, form bound states called hadrons, such as the proton, neutron, or B mesons. The decay channel of the B meson studied in this thesis is $B^0 \to \phi K^*$, the decay of a B^0 meson into a ϕ and K^* meson. This decay is illustrated in Fig. 2.2, by a Feynman diagram. Each line and vertex in a Feynman diagram is directly related to the underlying theory. Feynman diagrams provide understanding of a process without the necessity of doing all mathematical calculations.

A detailed and more mathematical introduction into the SM of particle physics and quantum field theory can be found in Ref. [15] or many other textbooks. Particle properties are regularly summarized and updated by the particle data group in the review of particle physics [16] which also includes a summary of the theoretical concepts [16, Chaps. 9 and 10].

2.2 The CKM Matrix

The top left vertex in Fig. 2.2, related to the $\bar{b} \to \bar{u}, \bar{c}, \bar{t}$ transition under emission of a W^+ boson, is a flavor-changing charged-current process. Such a transition is possible as the mass eigenstates of the quarks, as shown in Fig. 2.1, do not coincide with the weak eigenstates of the quarks. Otherwise, only transitions within the same generation of quarks would be possible. In the SM, this is realized by a rotation matrix for the down-type quarks. This principle was first introduced by Cabibbo [17], for two generations of quarks, and later extended by Kobayashi and Maskawa [18] to three generations. The matrix is know as the CKM matrix and relates the weak eigenstates of the down-type quarks to their mass eigenstates by

$$\begin{pmatrix} d \\ s \\ b \end{pmatrix}_{\text{weak}} = \begin{pmatrix} V_{ud} & V_{us} & V_{ub} \\ V_{cd} & V_{cs} & V_{cb} \\ V_{td} & V_{ts} & V_{tb} \end{pmatrix} \begin{pmatrix} d \\ s \\ b \end{pmatrix}_{\text{mass}} \tag{2.3}$$

The CKM matrix is complex and supposed to be unitary if there are no more than three generations of quarks. With three generations, there exists an irreducible complex phase in this matrix. The CKM element V_{ij} appears at every charged current vertex with i and j being the flavor of the related quarks, whereas the complex-conjugated element V_{ij}^* appears at the corresponding vertex of the *CP*-conjugated process. Without the irreducible complex phase $V_{ij} = V_{ij}^*$ would be satisfied for all elements. The phase leads to $V_{ij} \neq V_{ij}^*$ and is the only mechanism in the SM to explain *CP* violation. Kobayashi and Maskawa realized this long before the third generation was known.

There exist different parametrizations of the CKM matrix, of which the Wolfenstein parametrization [19] is the most natural one. This parametrization provides a representation for the hierarchical structure of the CKM matrix and the irreducible complex phase is related to the elements V_{td} and V_{ub}. Therefore, a sizeable *CP* violation in transitions involving b quarks was expected. After the phase was measured to be non-zero by Belle [20] and BaBar [21] in 2001, *CP* violation in the neutral B meson system was established; resulting in the Nobel prize for Kobayashi and Maskawa in 2008. This mechanism is however not strong enough to explain the observed matter-antimatter asymmetry in the universe.

A more detailed review of the CKM matrix formalism, the Wolfenstein parametrization, and experimental results on the determination of the individual matrix elements can be found in Ref. [16, Chap. 11].

2.3 CP Violation

The complex phase in the CKM matrix can manifest itself in three different types of *CP* violation: *CP* violation in the decay also called direct *CP* violation, *CP* violation in mixing, and *CP* violation in the inference between mixing and decay or mixing-induced *CP* violation. The latter two, as their name indicates, involve neutral meson mixing and are not studied in this thesis.

Direct *CP* violation is the only source of *CP* violation in charged meson decays and can also occur in neutral meson decays. It occurs if the total amplitudes of the *CP*-conjugated processes are different, i. e.

$$\left| \bar{A}_{\bar{f}} / A_f \right| \neq 1, \tag{2.4}$$

where f $(\bar{f})$ is a label for the transition from initial to final state $I \to f$ $(\bar{I} \to \bar{f})$. The total amplitude of such transitions, as they are studied in decay processes, can be written as a coherent sum of the contributing decay amplitudes:

$$A_f = \sum_j |a_j| \, e^{i(\delta_j + \phi_j)} \tag{2.5}$$

and

$$\bar{A}_{\bar{f}} = \sum_j |a_j| \, e^{i(\delta_j - \phi_j)}, \tag{2.6}$$

where j runs over the contributing amplitudes with magnitude a_j, δ_j is an associated *CP* conserving phase of the amplitude, and ϕ_j an associated *CP* violating phase that enters with opposite sign into A_f and $\bar{A}_{\bar{f}}$. An asymmetry related to *CP* violation can be defined as

$$\mathcal{A}_{CP} = \frac{\Gamma_{\bar{f}} - \Gamma_f}{\Gamma_{\bar{f}} + \Gamma_f} = \frac{\left|\bar{A}_{\bar{f}}/A_f\right| - 1}{\left|\bar{A}_{\bar{f}}/A_f\right| + 1} \propto 2 \sum_{i,j} |a_i a_j| \sin(\delta_i - \delta_j) \sin(\phi_i - \phi_j), \tag{2.7}$$

where Γ_f is the decay width and f the same label as before, $\delta_i - \delta_j$ is known as the strong phase difference between two contributing amplitudes, and $\phi_i - \phi_j$ as the weak phase difference. The weak phase difference originates from the complex phase of the CKM matrix elements which enters with opposite sign into A_f and $\bar{A}_{\bar{f}}$. The strong phase difference originates from hadronic effects, e. g. from rescattering of intermediate on-shell states. Since the strong interaction is invariant under *CP* transformation, the strong phase is equal for A_f and $\bar{A}_{\bar{f}}$. From Eq. (2.7) the three necessary conditions for direct *CP* violation can be seen; at least two contributing decay amplitudes with non-vanishing strong and weak phase differences.

In the SM, negligible *CP* violation is expected in charmless $B^0 \to \phi K^*$ decays. The decay is dominated by the amplitude of the loop process shown in Fig. 2.2. In models beyond the SM, new particles could appear in virtual loops in the Feynman diagram. Such new contributing amplitudes could result in significant deviations from the SM expectation and therefore can provide answers to cosmological problems.

A more detailed review on the *CP* violation formalism, the other two types of *CP* violation, the related neutral meson mixing formalism, and an overview of experimental results can be found in Ref. [16, Chap. 12].

2.4 Polarization in B Meson Decays

Depending on the decay channel of a B meson, e. g. in decays to two vector mesons (vector–vector) or a vector and a tensor meson (vector–tensor), the daughter particles might be polarized and not decay isotropically. The polarization of such decays is related to the strong interaction. From polarization studies one may therefore gain a deeper understanding of QCD. The naive expectation based on the factorization approach [22] predicts a longitudinal polarization fraction f_L close to unity for charmless B meson decays. An overview of the latest results on longitudinal

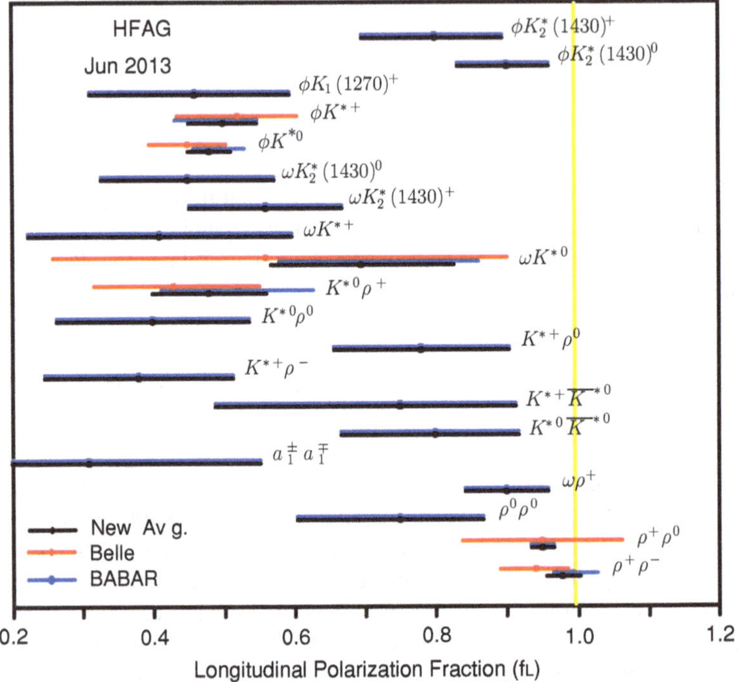

Fig. 2.3 Longitudinal polarization fraction in different charmless B meson decays as of June 2013. Provided by the Heavy Flavor Averaging Group (HFAG), see also Ref. [23]

polarization fractions in charmless B meson decays is shown in Fig. 2.3. For many decay modes, the measured polarization fraction deviates from the naive expectation.

Of particular interest for this work are the measurements of the longitudinal polarization fraction in $B^0 \to \phi K^*(892)^0$ decays. The Belle and BaBar collaboration measured it to be $f_L = 0.45 \pm 0.05 \pm 0.02$ [24] and $f_L = 0.494 \pm 0.034 \pm 0.013$ [25], respectively. These measurements deviate from the naive expectation, whereas BaBar measured $f_L = 0.901^{+0.046}_{-0.058} \pm 0.037$ [25] in $B^0 \to \phi K_2^*(1430)^0$ decays, which is consistent with the factorization approach. One aim of the analysis presented in this thesis is to provide an improved measurement of polarization in $B^0 \to \phi K^*$ decays, including $B^0 \to \phi K_2^*(1430)^0$ decays, with data from the Belle experiment.

Polarization measurements, using flavor-specific $B^0 \to \phi K^*$ decays, can further be used to distinguish the *CP*-even and -odd fraction in $B^0/\overline{B}^0 \to \phi(K_S^0\pi^0)^*$ decays. This decay channel can be used for time-dependent measurements of mixing-induced *CP* violation in $b \to (s\bar{s})s$ transitions, which is beyond the scope of this thesis.

A more detailed review on polarization in B meson decays can be found in Refs. [26] and [22]. A detailed description for $B^0 \to \phi K^*$ decays will also be derived in the subsequent section.

2.5 Partial Wave Analysis

The technique of partial wave analysis is employed to measure the polarization in the decay $B^0 \to \phi K^*$. The flavor-specific decay $B^0 \to \phi K^*$ with $\phi \to K^+K^-$ and $K^* \to K^+\pi^-$ is used, where K^* denotes contributions from scalar (S-wave, spin $J = 0$), vector (P-wave, $J = 1$), and tensor (D-wave, $J = 2$) components from $(K\pi)^*_0$, $K^*(892)^0$, and $K^*_2(1430)^0$, respectively. As the ϕ meson is a vector meson, the pseudoscalar B^0 decays into a vector–scalar, vector–vector, and vector–tensor state for S-, P-, and D-wave, respectively. The partial wave analysis uses the different mass and angular distributions of the three contributing channels $B^0 \to \phi(K\pi)^*_0$, $B^0 \to \phi K^*(892)^0$, and $B^0 \to \phi K^*_2(1430)^0$ to distinguish among them and determine the polarization in the vector–vector and vector–tensor decay.

The polarization in the two flavor-specific decays $B^0 \to \phi K^*$ and $\overline{B}^0 \to \phi(K^-\pi^+)^*$ is measured simultaneously to determine a number of parameters related to direct *CP* violation. Throughout this thesis, the inclusion of the charged-conjugated mode is implied unless otherwise stated.

The $K^+\pi^-$ invariant mass is studied below 1.55 GeV, as the LASS model [27], used to parametrize the S-wave contribution and described below, is not valid above this value. Furthermore, no significant contribution from K^* states beyond 1.55 GeV is observed [28].

Following, the helicity formalism is introduced to describe angular distributions. Furthermore, the parametrization of the $K^+\pi^-$ invariant-mass distribution for the S-, P-, and D-wave is described. Finally, the combined model of mass and angular distribution of partial waves, which is applied for the parameter extraction, is derived.

2.5.1 Angular Distribution

The angular distribution in the $B^0 \to \phi K^*$ system with $\phi \to K^+K^-$ and $K^* \to K^+\pi^-$ is described by the three helicity angles θ_1, θ_2, and Φ, which are defined in the rest frame of the parent particles as illustrated in Fig. 2.4. The angle Φ is defined as the angle between the decay planes of the K^* and ϕ meson in the Brest frame. The angle θ_1 (θ_2) is defined as the angle between the direction of the K^* (ϕ) meson and the K^+ daughter in the K^* (ϕ) rest frame.

As derived in Ref. [29], due to angular momentum conservation, the partial decay width for a two-body decay of a pseudoscalar B meson into particles with spins J_1 and J_2 is given by

$$\frac{d^3\Gamma}{d\cos\theta_1 d\cos\theta_2 d\Phi} \propto \left|\sum_\lambda A_\lambda Y^\lambda_{J_1}(\theta_1, \Phi)\, Y^{-\lambda}_{J_2}(-\theta_2, 0)\right|^2, \tag{2.8}$$

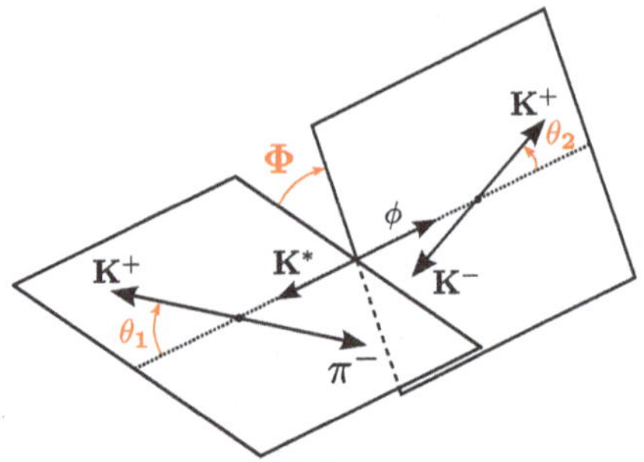

Fig. 2.4 Definition of the three helicity angles given in the rest frame of the parent particles for the $B^0 \to \phi K^*$ decay

where Y_l^m are the spherical harmonics, the sum is over the helicity states λ, and A_λ is the complex weight of the corresponding helicity amplitude. The parameter λ takes all discrete values between $-j$ and $+j$, with j being the smaller of the two daughter particle spins J_1 and J_2. As the ϕ is a vector meson, $J_2 = 1$ in the following, whereas $J_1 = 0$ for $(K\pi)_0^*$, $J_1 = 1$ for $K^*(892)^0$, and $J_1 = 2$ for $K_2^*(1430)^0$. The partial decay width of each partial wave with spin $J \equiv J_1$ is therefore

$$\frac{d^3\Gamma}{d\cos\theta_1 d\cos\theta_2 d\Phi} \propto \left| \sum_\lambda A_{J\lambda} Y_J^\lambda(\theta_1, \Phi)\, Y_1^{-\lambda}(-\theta_2, 0) \right|^2, \tag{2.9}$$

with $A_{J\lambda}$ being the complex weight of the corresponding helicity amplitude of the partial wave with spin J.

The helicity basis is not a basis of *CP* eigenstates. Polarization measurements are often performed in the transversity basis [26] of *CP* eigenstates with the transformation $A_{J\pm1} = (A_{J\parallel} \pm A_{J\perp})/\sqrt{2}$ for two of the amplitudes. In this basis, the longitudinal polarization A_{J0} and the parallel polarization $A_{J\parallel}$ are even under *CP* transformation while the perpendicular component $A_{J\perp}$ is *CP*-odd. Throughout this thesis, A is used for B^0 and $\bar{A}$ for $\bar{B}^0$ related complex weights of the helicity and transversity amplitudes. Furthermore, depending on the context, either of the two bases is used with $\lambda = -1, 0, +1$ or $\lambda = 0, \parallel, \perp$. Where necessary, the basis used is explicitly stated. The complex weights are defined using polar coordinates $A_{J\lambda} = a_{J\lambda}e^{i\varphi_{J\lambda}}$ and apply the same implicit definition of the basis; e.g. $a_{2\perp}$ would be the magnitude of the perpendicular D-wave component in the transversity basis.

2.5.2 Mass Distribution

The $K^+\pi^-$ invariant-mass spectrum $M_{K\pi}$ is studied to distinguish among different partial waves. To parametrize the lineshape of the P- and D-wave components as a function of the invariant mass m, a relativistic spin-dependent Breit–Wigner (BW) amplitude R_J [30] is used:

$$R_J(m) = \frac{m_J \Gamma_J(m)}{(m_J^2 - m^2) - i m_J \Gamma_J(m)} = \sin\delta_J e^{i\delta_J}, \tag{2.10}$$

where the convention

$$\cot\delta_J = \frac{m_J^2 - m^2}{m_J \Gamma_J(m)} \tag{2.11}$$

is applied. For spin $J = 1$ and $J = 2$, the mass-dependent widths are given by

$$\Gamma_1(m) = \Gamma_1 \frac{m_1}{m} \frac{1 + r^2 q_1^2}{1 + r^2 q^2} \left(\frac{q}{q_1}\right)^3, \tag{2.12}$$

$$\Gamma_2(m) = \Gamma_2 \frac{m_2}{m} \frac{9 + 3r^2 q_2^2 + r^4 q_2^4}{9 + 3r^2 q^2 + r^4 q^4} \left(\frac{q}{q_2}\right)^5, \tag{2.13}$$

where Γ_J is the resonance width, m_J the resonance mass, q the momentum of a daughter particle in the rest frame of the resonance, q_J this momentum evaluated at $m = m_J$, and r the interaction radius. This parametrization of the mass-dependent width uses the Blatt–Weisskopf penetration factors [30].

The S-wave component is parametrized using Kπ scattering results from the LASS experiment [27]. It was found by LASS that the scattering is elastic up to about 1.5–1.6 GeV and thus can be parametrized as

$$R_0(m) = \sin\delta_0 e^{i\delta_0}, \tag{2.14}$$

where

$$\delta_0 = \Delta R + \Delta B, \tag{2.15}$$

ΔR is representing a resonant contribution from $K_0^*(1430)^0$ while ΔB is denoting a non-resonant contribution. The resonant part is defined as

$$\cot\Delta R = \frac{m_0^2 - m^2}{m_0 \Gamma_0(m)}, \tag{2.16}$$

where m_0 and Γ_0 are the resonance mass and width, and $\Gamma_0(m)$ is given by

$$\Gamma_0(m) = \Gamma_0 \frac{m_0}{m} \left(\frac{q}{q_0}\right). \tag{2.17}$$

The non-resonant part is defined as

$$\cot\Delta B = \frac{1}{aq} + \frac{bq}{2}, \tag{2.18}$$

where a is the scattering length and b is the effective range.

The amplitude $M_J(m)$ for the partial wave with spin J is obtained by multiplying the lineshape with the two-body phase space factor

$$M_J(m) = \frac{m}{q} R_J(m). \tag{2.19}$$

The K^+K^- invariant-mass spectrum M_{KK} is the same for the three contributing partial waves. The ϕ lineshape is parametrized by a relativistic spin-dependent BW with spin $J = 1$. In the following, the explicit dependence of the mass-angular distribution on the M_{KK} distribution is omitted. In the analysis itself, the M_{KK} distribution is taken into account and details are discussed in Chap. 6.

2.5.3 *Mass-Angular Distribution*

The mass distribution is combined with the angular distribution to obtain the partial decay width

$$\frac{d^4\Gamma}{d\cos\theta_1 d\cos\theta_2 d\Phi d M_{K\pi}} \propto |\mathcal{M}\left(M_{K\pi}, \cos\theta_1, \cos\theta_2, \Phi\right)|^2 \times F_{M_{\phi K}}\left(M_{K\pi}\right), \tag{2.20}$$

where $F_{M_{\phi K}}(M_{K\pi})$ is a phase space factor that takes into account the three-body kinematics in $B^0 \to \phi K^+\pi^-$. As no resonant charmless structure is expected in the ϕK^+ invariant-mass distribution, a constant amplitude is assumed in the ϕK^+ invariant mass $M_{\phi K}$ that can be computed for each value of $M_{K\pi}$ following the section on kinematics in Ref. [16] as

$$F(m) = 2m\left[m^2_{\max}(m) - m^2_{\min}(m)\right], \tag{2.21}$$

with $m^2_{\max}$ ($m^2_{\min}$) being the maximum (minimum) value of the Dalitz plot range of the ϕK^+ invariant mass $M_{\phi K}$ at a given $M_{K\pi}$ value m.

The matrix element squared $|\mathcal{M}(M_{K\pi}, \cos\theta_1, \cos\theta_2, \Phi)|^2$ is given by the coherent sum of the corresponding S-, P-, and D-wave amplitudes A_J as

$$\begin{aligned} |\mathcal{M}\left(M_{K\pi}, \cos\theta_1, \cos\theta_2, \Phi\right)|^2 = |&\mathcal{A}_0\left(M_{K\pi}, \cos\theta_1, \cos\theta_2, \Phi\right) \\ &+ \mathcal{A}_1\left(M_{K\pi}, \cos\theta_1, \cos\theta_2, \Phi\right) \\ &+ \mathcal{A}_2\left(M_{K\pi}, \cos\theta_1, \cos\theta_2, \Phi\right)|^2, \end{aligned} \tag{2.22}$$

where the explicit dependence of $\mathcal{M}$ on $(M_{K\pi}, \cos\theta_1, \cos\theta_2, \Phi)$ is omitted for readability in the following. Each partial wave for a given spin J is parametrized as the product of the angular distribution from Eq. (2.9) and the mass distribution from Eq. (2.19). For the S-, P-, and D-wave,

$$\mathcal{A}_0\,(M_{K\pi}, \cos\theta_1, \cos\theta_2, \Phi) = A_{00} Y_0^0(\theta_1, \Phi) Y_1^0(-\theta_2, 0) \times M_0(M_{K\pi}), \quad (2.23)$$

$$\mathcal{A}_1\,(M_{K\pi}, \cos\theta_1, \cos\theta_2, \Phi) = \sum_{\lambda=0,\pm1} A_{1\lambda} Y_1^{\lambda}(\theta_1, \Phi) Y_1^{-\lambda}(-\theta_2, 0) \times M_1(M_{K\pi}), \quad (2.24)$$

and

$$\mathcal{A}_2\,(M_{K\pi}, \cos\theta_1, \cos\theta_2, \Phi) = \sum_{\lambda=0,\pm1} A_{2\lambda} Y_2^{\lambda}(\theta_1, \Phi) Y_1^{-\lambda}(-\theta_2, 0) \times M_2(M_{K\pi}) \quad (2.25)$$

is obtained, respectively. Overall, the seven complex helicity amplitudes contributing to these formulas can be parametrized by 14 real parameters (28 if B^0 and $\overline{B}^0$ are measured independently).

The normalized partial decay width can be defined as

$$\frac{d^4\Gamma}{d\cos\theta_1 d\cos\theta_2 d\Phi dM_{K\pi}} = \frac{(1+Q)\times|\mathcal{M}^+|^2 + (1-Q)\times|\mathcal{M}^-|^2}{2\mathcal{N}} \times F_{M_{\phi K}}\,(M_{K\pi}), \quad (2.26)$$

where $\mathcal{M}^+$ [$\mathcal{M}^-$] is the matrix element for $\overline{B}^0 \to \phi(K^-\pi^+)^*$ [$\overline{B}^0 \to \phi(K^-\pi^+)^*$], Q is ±1 depending on the charge of the primary charged kaon from the B meson, and $\mathcal{N}$ is the overall normalization given by

$$\mathcal{N} = \frac{1}{2}\int |\mathcal{M}^+|^2 \times F_{M_{\phi K}}\,(M_{K\pi})\, d\cos\theta_1 d\cos\theta_2 d\Phi dM_{K\pi}$$
$$+ \frac{1}{2}\int |\mathcal{M}^-|^2 \times F_{M_{\phi K}}\,(M_{K\pi})\, d\cos\theta_1 d\cos\theta_2 d\Phi dM_{K\pi}. \quad (2.27)$$

By averaging the normalization over B^0 and $\overline{B}^0$, a simultaneous fit with a single reference amplitude of fixed magnitude, which defines the relative strengths of the amplitudes, can be performed. If both final states are normalized independently, each with its own reference amplitude, and *CP* violation is observed, the interpretation of whether *CP* violation is in the reference amplitudes or all other amplitudes would be ambiguous.

Using these notations, the final set of parameters used in the analysis presented in this thesis can be defined. For the matrix element $\mathcal{M}^+$, the weights are defined as $A_{J\lambda} = a^+_{J\lambda} e^{i\varphi^+_{J\lambda}}$ and, for $\mathcal{M}^-$, as $\bar{A}_{J\lambda} = a^-_{J\lambda} e^{i\varphi^-_{J\lambda}}$. Here $a^\pm_{J\lambda}$ is defined as

$$a^\pm_{J\lambda} = a_{J\lambda}(1 \pm \Delta a_{J\lambda}) \quad (2.28)$$

and $\varphi^\pm_{J\lambda}$ is given by

$$\varphi^\pm_{J\lambda} = \varphi_{J\lambda} \pm \Delta\varphi_{J\lambda}, \quad (2.29)$$

Table 2.1 Definitions of the 26 real parameters that are measured in the $B^0 \to \phi K^*$ system

Parameter	Definition	$\phi(K\pi)^*_0$ $J=0$	$\phi K^*(892)^0$ $J=1$	$\phi K^*_2(1430)^0$ $J=2$
$\mathcal{B}_J$	$\frac{1}{2}(\bar{\Gamma}_J + \Gamma_J)/\Gamma_{total}$	$\mathcal{B}_0$	$\mathcal{B}_1$	$\mathcal{B}_2$
f_{LJ}	$\frac{1}{2}(\lvert\bar{A}_{J0}\rvert^2/\sum\lvert\bar{A}_{J\lambda}\rvert^2 + \lvert A_{J0}\rvert^2/\sum\lvert A_{J\lambda}\rvert^2)$	...	f_{L1}	f_{L2}
$f_{\perp J}$	$\frac{1}{2}(\lvert\bar{A}_{J\perp}\rvert^2/\sum\lvert\bar{A}_{J\lambda}\rvert^2 + \lvert A_{J\perp}\rvert^2/\sum\lvert A_{J\lambda}\rvert^2)$	...	$f_{\perp 1}$	$f_{\perp 2}$
$\phi_{\parallel J}$	$\frac{1}{2}(\arg(\bar{A}_{J\parallel}/\bar{A}_{J0}) + \arg(A_{J\parallel}/A_{J0}))$	...	$\phi_{\parallel 1}$	$\phi_{\parallel 2}$
$\phi_{\perp J}$	$\frac{1}{2}(\arg(\bar{A}_{J\perp}/\bar{A}_{J0}) + \arg(A_{J\perp}/A_{J0}) - \pi)$	...	$\phi_{\perp 1}$	$\phi_{\perp 2}$
δ_{0J}	$\frac{1}{2}(\arg(\bar{A}_{00}/\bar{A}_{J0}) + \arg(A_{00}/A_{J0}))$	...	δ_{01}	δ_{02}
$\mathcal{A}_{CPJ}$	$(\bar{\Gamma}_J - \Gamma_J)/(\bar{\Gamma}_J + \Gamma_J)$	$\mathcal{A}_{CP0}$	$\mathcal{A}_{CP1}$	$\mathcal{A}_{CP2}$
$\mathcal{A}^0_{CPJ}$	$\frac{\lvert\bar{A}_{J0}\rvert^2/\sum\lvert\bar{A}_{J\lambda}\rvert^2 - \lvert A_{J0}\rvert^2/\sum\lvert A_{J\lambda}\rvert^2}{\lvert\bar{A}_{J0}\rvert^2/\sum\lvert\bar{A}_{J\lambda}\rvert^2 + \lvert A_{J0}\rvert^2/\sum\lvert A_{J\lambda}\rvert^2}$	...	$\mathcal{A}^0_{CP1}$	$\mathcal{A}^0_{CP2}$
$\mathcal{A}^\perp_{CPJ}$	$\frac{\lvert\bar{A}_{J\perp}\rvert^2/\sum\lvert\bar{A}_{J\lambda}\rvert^2 - \lvert A_{J\perp}\rvert^2/\sum\lvert A_{J\lambda}\rvert^2}{\lvert\bar{A}_{J\perp}\rvert^2/\sum\lvert\bar{A}_{J\lambda}\rvert^2 + \lvert A_{J\perp}\rvert^2/\sum\lvert A_{J\lambda}\rvert^2}$	...	$\mathcal{A}^\perp_{CP1}$	$\mathcal{A}^\perp_{CP2}$
$\Delta\phi_{\parallel J}$	$\frac{1}{2}(\arg(\bar{A}_{J\parallel}/\bar{A}_{J0}) - \arg(A_{J\parallel}/A_{J0}))$	...	$\Delta\phi_{\parallel 1}$	$\Delta\phi_{\parallel 2}$
$\Delta\phi_{\perp J}$	$\frac{1}{2}(\arg(\bar{A}_{J\perp}/\bar{A}_{J0}) - \arg(A_{J\perp}/A_{J0}) - \pi)$	...	$\Delta\phi_{\perp 1}$	$\Delta\phi_{\perp 2}$
$\Delta\delta_{0J}$	$\frac{1}{2}(\arg(\bar{A}_{00}/\bar{A}_{J0}) - \arg(A_{00}/A_{J0}))$	...	$\Delta\delta_{01}$	$\Delta\delta_{02}$

Three partial waves with spin $J = 0, 1, 2$ are considered in the $K^+\pi^-$ spectrum. The amplitude weights $A_{J\lambda}$ and $\bar{A}_{J\lambda}$ are defined in the text. The extra π in the definition of $\phi_{\perp J}$ and $\Delta\phi_{\perp J}$ accounts for the sign flip of $A_{J\perp} = -\bar{A}_{J\perp}$ under *CP* transformation

where one *CP*-conserving and one *CP*-violating parameter is used per magnitude and phase. For $J = 0$ only $\lambda = 0$ is possible, whereas, for $J = 1$ and $J = 2$, three values $\lambda = 0$, $\parallel$ and $\perp$ are allowed.

As reference phase $\varphi_{00} = 0$ is chosen, as the system is invariant under a global phase transformation. This effectively reduces the 28 parameters by one. Of the remaining 27 parameters, 26 can be measured in the $B^0 \to \phi K^*$ system with $K^* \to K^+\pi^-$. These 26 parameters can be used to define a more common set of parameters shown in Table 2.1, which are used in the review of polarization in B decays in Ref. [26]. For each partial wave J, parameters such as the longitudinal (perpendicular) polarization fractions f_{LJ} ($f_{\perp J}$), the relative phase of the parallel (perpendicular) amplitude $\phi_{\parallel J}$ ($\phi_{\perp J}$) to the longitudinal amplitude, and the strong phase difference between the partial waves δ_{0J} and a number of parameters related to *CP* violation are defined. The 27th parameter, $\Delta\varphi_{00} = \Delta\phi_{00} = \frac{1}{2}\arg(A_{00}/\bar{A}_{00})$, could only be measured in a time-dependent analysis of *CP* violation in $B^0/\bar{B}^0 \to \phi(K^0_S\pi^0)^*$ decays that is beyond the scope of this thesis, so $\Delta\varphi_{00}$ is fixed to zero. Furthermore, a_{10} is fixed as it has the largest relative magnitude among all amplitudes and chosen as the reference amplitude. Fixing a_{10} does not decrease the number of free parameters as the absolute magnitude, defined by the signal yield, remains a free parameter in the fit. Overall, 26 real parameters are left to be determined.

In the previous Belle analysis [24], a twofold phase ambiguity was observed in the decay of $B^0 \to \phi K^*(892)^0$; this is a fourfold ambiguity if B^0 and $\bar{B}^0$ are measured independently, as the sets $(\phi_{\parallel J}, \phi_{\perp J}, \Delta\phi_{\parallel J}, \Delta\phi_{\perp J})$ and $(2\pi - \phi_{\parallel J}, \pi - \phi_{\perp J}, -\Delta\phi_{\parallel J}, -\Delta\phi_{\perp J})$ solve all angular equations. Even the interference terms in

$|\mathcal{M}|^2$ are invariant under such transformation if the sign of the strong phase δ_{0J} is flipped. However, the mass dependence of δ_{0J} is unique: it either increases or decreases with increasing $K^+\pi^-$ invariant mass. The ambiguity is solved for B^0 and $\overline{B}^0$ using Wigner's causality principle [31], which states that the phase of a resonance increases with increasing invariant mass.

2.5.4 Triple-Product Correlations

From the measured weights $A_{J\lambda}$, one can also calculate the triple-product correlations. These quantities have been given in the previous Belle analysis [24] and are used in other polarization measurements. The triple-product correlations do not contain additional information with respect to Table 2.1, but are an alternative representation. The T-odd quantities

$$\overset{(-)}{A}{}^{0}_{T,J} = \mathrm{Im}(\overset{(-)}{A}_{J\perp}\overset{(-)}{A}{}^{*}_{J0}) \tag{2.30}$$

and

$$\overset{(-)}{A}{}^{\parallel}_{T,J} = \mathrm{Im}(\overset{(-)}{A}_{J\perp}\overset{(-)}{A}{}^{*}_{J\parallel}) \tag{2.31}$$

from Ref. [32] and the corresponding asymmetries $\mathcal{A}^{0/\parallel}_{T,J}$ between B^0 and $\overline{B}^0$ are sensitive to T-odd CP violation in a given decay channel with spin J.

References

1. S.L. Glashow, Partial-symmetries of weak interactions. Nucl. Phys. **22**(4), 579–588 (1961)
2. S. Weinberg, A model of leptons. Phys. Rev. Lett. **19**(21), 1264–1266 (1967)
3. A. Salam, *Elementary Particle Theory* (Almquist & Wiksells, Stockholm, 1969)
4. D.J. Gross, F. Wilczek, Ultraviolet behavior of non-abelian gauge theories. Phys. Rev. Lett. **30**(26), 1343–1346 (1973)
5. H.D. Politzer, Reliable perturbative results for strong interactions. Phys. Rev. Lett. **30**(26), 1346–1349 (1973)
6. D.J. Gross, F. Wilczek, Asymptotically free gauge theories. I. Phys. Rev. D **8**(10), 3633–3652 (1973)
7. H.D. Politzer, Asymptotic freedom: an approach to strong interactions. Phys. Rep. **14**, 129–180 (1974)
8. Figure from the Wikimedia Commons and licensed under the Creative Commons Attribution 3.0 Unported licence. Online at http://en.wikipedia.org/wiki/File:Standard_Model_of_Elementary_Particles.svg. Accessed 24 July 2013
9. P. Higgs, Broken symmetries, massless particles and gauge fields. Phys. Lett. **12**(2), 132–133 (1964)
10. The ATLAS Colloboration, Observation of a new particle in the search for the Standard Model Higgs boson with the ATLAS detector at the LHC. Phys. Lett. B **716**(1), 1–29 (2012)

11. The CMS Collaboration, Observation of a new boson at a mass of 125 GeV with the CMS experiment at the LHC. Phys. Lett. B **716**(1), 30–61 (2012)
12. G. Arnison et al., (UA1 Collaboration). Experimental observation of isolated large transverse energy electrons with associated missing energy at $\sqrt{s} = 540$ GeV. Phys. Lett. B **122**(1), 103–116 (1983)
13. M. Banner et al., (UA2 Collaboration). Observation of single isolated electrons of high transverse momentum in events with missing transverse energy at the CERN $\bar{p}p$ collider. Phys. Lett. B **122**(5–6), 476–485 (1983)
14. G. Arnison et al., (UA1 Collaboration). Experimental observation of lepton pairs of invariant mass around [95]GeV/c^2 at the CERN SPS collider. Phys. Lett. B **125**(5), 398–410 (1983)
15. L. Alvarez-Gaume, M.A. Vazquez-Mozo, *An Invitation to Quantum Field Theory, Volume 839 of Lecture Notes in Physics* (Springer, New York, 2011)
16. J. Beringer et al., (Particle Data Group). The review of particle physics. Phys. Rev. D **86**, 010001 (2012). Online version available at http://pdg.lbl.gov
17. N. Cabibbo, Unitary symmetry and leptonic decays. Phys. Rev. Lett. **10**, 531–533 (1963)
18. M. Kobayashi, T. Maskawa, CP violation in the renormalizable theory of weak interaction. Prog. Theor. Phys. **49**, 652–657 (1973)
19. L. Wolfenstein, Parametrization of the Kobayashi-Maskawa matrix. Phys. Rev. Lett. **51**(21), 1945–1947 (1983)
20. K. Abe et al., (Belle Collaboration). Observation of large CP violation in the neutral B meson system. Phys. Rev. Lett. **87**, 091802 (2001)
21. B. Aubert et al., (BaBar Collaboration). Observation of CP violation in the B^{0}meson system. Phys. Rev. Lett. **87**, 091801 (2001)
22. A. Datta et al., Study of polarization in B $\rightarrow$ VT decays. Phys. Rev. D **77**, 114025 (2008). See also references therein for discussions of SM and new-physics explanations of f_T/f_L.
23. Y. Amhis et al., (Heavy Flavor Averaging Group). Averages of b-hadron, c-hadron, and tau-lepton properties as of early 2012. `arXiv:hep-ex/1207.1158` and online update at http://www.slac.stanford.edu/xorg/hfag
24. K.-F. Chen et al., (Belle Collaboration). Measurement of polarization and triple-product correlations in B $\rightarrow \phi$K* decays. Phys. Rev. Lett. **94**, 221804 (2005)
25. B. Aubert et al., (BaBar Collaboration). Time-dependent and time-intergrated angular analysis of B $\rightarrow \phi$K$_0^S\pi^0$ and ϕK$^\pm\pi^\pm$. Phys. Rev. D **78**, 092008 (2008)
26. J. Beringer et al., (Particle Data Group). Polarization in B decays. Review article available as part of the online version of Reference [16]
27. D. Aston et al., (LASS Collaboration). A study of K$^-\pi^+$ scattering in the reaction K$^-$p $\rightarrow$ K$^-\pi^+$ at 11 GeV/c. Nucl. Phys. B **296**, 493–526 (1988)
28. B. Aubert et al., (BaBar Collaboration). Search for B $\rightarrow \phi$(K$^+\pi^-$) decays with large K$^+\pi^-$ invariant mass. Phys. Rev. D **76**, 051103(R) (2007)
29. M. Jacob, G.C. Wick, On the general theory of collisions for particles with spin. Ann. Phy. **7**(4), 404–428 (1959)
30. J. Beringer et al., (Particle Data Group). Dalitz plot analysis formalism. Review article available as part of the online version of Reference [16]
31. E.P. Wigner, Lower limit for the energy derivative of the scattering phase shift. Phys. Rev. **98**, 145–147 (1955)
32. A. Datta, D. London, Triple-product correlations in B $\rightarrow$ V$_1$V$_2$ decays and new physics. Int. J. Mod. Phys. A **19**, 2505–2544 (2004). The definition $A_T^0 = A_T^{(1)}$ and $A_T^{\parallel} = A_T^{(2)}$ is taken, but $\bar{A}_T^0 = -\bar{A}_T^{(1)}$ and $\bar{A}_T^{\parallel} = -\bar{A}_T^{(2)}$

Chapter 3
The Belle Experiment

This chapter is going to describe the technical setup without which this thesis would not have been possible. The KEKB accelerator and the Belle detector at the Japanese High Energy Accelerator Research Organisation (KEK) in Tsukuba, Japan, provide an excellent environment to study flavor physics and are both briefly described below.

3.1 KEKB Accelerator

The KEKB accelerator was an asymmetric-energy e^+e^- collider. It operated at a center-of-mass energy of $\sqrt{s} = 10.56\,\mathrm{GeV}$, which corresponds to the mass of the $\Upsilon(4S)$ resonance. The $\Upsilon(4S)$ resonance is a bound state of a b and $\overline{b}$ quark, with a mass about 20 MeV above the threshold for $B\overline{B}$ pair production. It decays almost exclusively into $B^0\overline{B}^0$ or B^+B^- meson pairs. Therefore, KEKB is called a B factory.

The accelerator consisted of two storage rings with a circumference of about 3 km and was located 11 m below surface. A schematic layout of the KEKB accelerator is shown in Fig. 3.1. The electrons (positrons) have been accelerated in a linear accelerator and stored in bunches in the high energy ring (HER) and low energy ring (LER) with an energy of 8 and 3.5 GeV, respectively. The storage rings had a single intersection point with a crossing angle of 22 Mrad. The bunches from HER and LER did collide in this interaction region (IR), around which the Belle detector was built.

The asymmetric energy of the KEKB accelerator results in a Lorentz boost of the center-of-mass system of $\beta\gamma = 0.425$ along the e^--beam direction in the laboratory reference system. The Lorentz boost causes a spatial separation of the decay vertices of the two B mesons from the $\Upsilon(4S)$ decay. Before decaying, both B mesons cover a short distance, their decay length, in the detector. The decay length difference Δz between the vertices gets increased by the boost and reaches experimentally accessible distances of about 200 μm, which can be translated into a decay time difference $\Delta t = \Delta z/\beta\gamma c$. One of the design goals of KEKB was to enable measurements of the decay time difference and allow for time-dependent measurements of B mesons.

M. Prim, *Polarization and CP Violation Measurements*, Springer Theses,
DOI: 10.1007/978-3-319-05756-9_3,

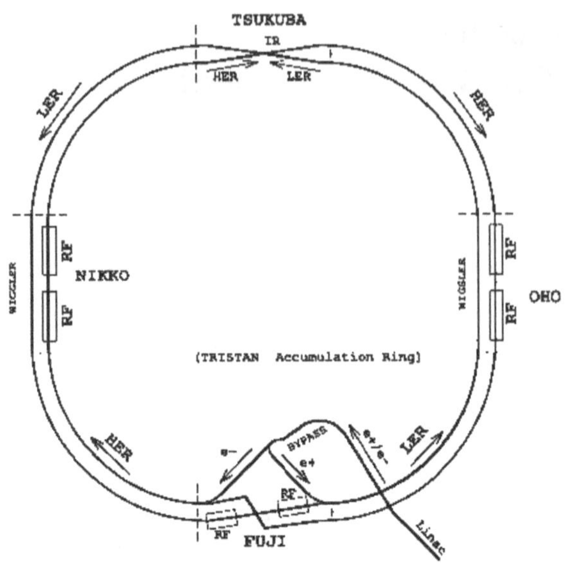

Fig. 3.1 Schematic layout of the KEKB accelerator. Taken from Ref. [1]

Another design goal of the KEKB accelerator was to achieve an instantaneous luminosity of $\mathcal{L} = 1.0 \times 10^{34}\,\mathrm{cm}^{-2}\,\mathrm{s}^{-1}$, which corresponds to a production rate of approximately 10 $\mathrm{B\overline{B}}$ pairs per second. The KEKB accelerator exceeded the design luminosity and did set various world records with peak luminosities as high as $\mathcal{L} = 2.1 \times 10^{34}\,\mathrm{cm}^{-2}\,\mathrm{s}^{-1}$. These records were achieved by improvements in the operation and accelerator design, such as the installation of crab cavities, special superconducting RF cavities that rotate the bunches to cause head-on collisions at the interaction point.

During the operation of KEKB as a B factory for the Belle detector from October 1999 to June 2010 more than $1\,\mathrm{ab}^{-1}$ of integrated luminosity was delivered. The Belle detector recorded $711\,\mathrm{fb}^{-1}$ at the $\Upsilon(4S)$ resonance. In addition, samples at different center-of-mass energies have been recorded. For example, $121\,\mathrm{fb}^{-1}$ have been recorded at the $\Upsilon(5S)$ resonance and $79\,\mathrm{fb}^{-1}$ below the $\Upsilon(4S)$.

Since its shutdown in 2010, KEKB is upgraded to the Super-B factory SuperKEKB with an up to 40 times increased instantaneous luminosity for the Belle II experiment. As of 2013, commissioning is planned for 2015.

A detailed description of the KEKB accelerator design and operation is given in Refs. [1–3].

3.2 Belle Detector

The Belle detector is a multi-purpose magnetic spectrometer built around the interaction region of the KEKB accelerator and covers a solid angle of 4π. It was initially designed for high precision time-dependent measurements but served also

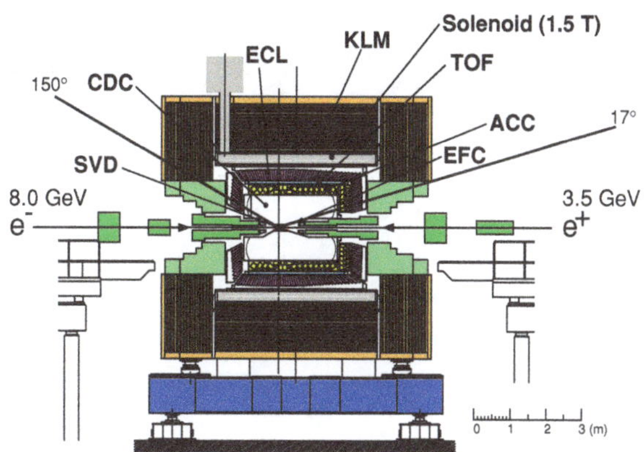

Fig. 3.2 Side view of the Belle detector with the different subdetectors. Taken from Ref. [4] and adapted

as an excellent place to study rare B meson decays, charm physics and flavor physics in general.

In Fig. 3.2 a schematic side view of the Belle detector is shown. The detector consists of several subdetectors to detect and identify charged and neutral particles. A superconducting solenoid provides a magnet field of 1.5 T. A silicon vertex detector (SVD) is located around the beam pipe and used for charged track and vertex reconstruction in combination with the central drift chamber (CDC). The measurement of dE/dx in the CDC, an array of aerogel Čerenkov counters (ACC) and an arrangement of time-of-flight scintillation counters (TOF) is used for particle identification of charged tracks. Electromagnetic showers are detected in the electromagnetic calorimeter (ECL) composed of CsI(Tl) crystals. An extreme forward calorimeter (EFC) out of BGO crystals is located close to the interaction region to increase the angular coverage and serves as a beam monitor and luminosity measurement device. Inside the iron support structure, which is used as yoke for the solenoid to return the magnetic flux and as absorber material, resistive plate counters are installed to detect K_L^0 mesons and identify muons (KLM). The signals from the different subdetectors are collected by a multi-level trigger and data acquisition (DAQ) system.

Since 2010, also the Belle detector is under a ongoing upgrade process. Nearly all detector components are removed and replaced by improved systems with better performance and radiation hardness, to keep up with the increased luminosity of SuperKEKB. As of 2013, commissioning of the Belle II [5] detector is planned for 2015.

In the next sections a brief description of the individual components of the Belle detector is given. The description starts from the innermost component and provides references to a series of technical publications that contain more detailed information. The coordinate system is chosen such that the z-axis points in direction of the electron beam. The polar angle θ is measured with respect to the z-axis and the $r\phi$-plane is chosen perpendicular to the z-axis.

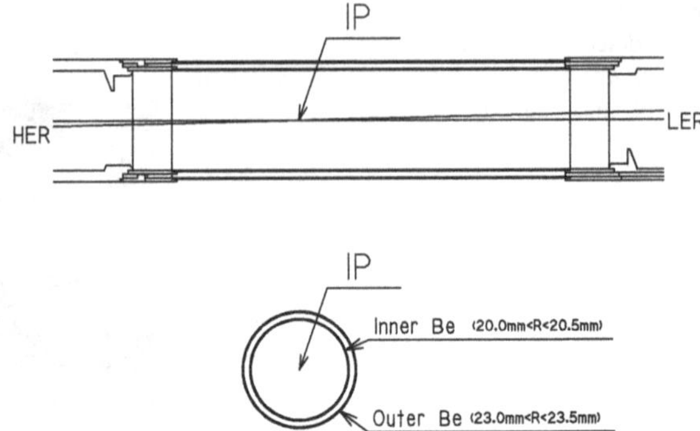

Fig. 3.3 Longitudinal and cross section of the beryllium beam pipe enclosing the interaction point as used from 1999 to 2003. Taken from Ref. [4]

3.2.1 Beam Pipe

The beam pipe surrounds the interaction point, the crossing of the high and low energy storage rings. It is designed as double-wall beryllium cylinder, see Fig. 3.3, with an inner radius of 20 mm and an outer radius of 23.5 mm. To reduce Coulomb scattering in the beam pipe, a limiting effect on the z-vertex resolution, each wall is only 0.5 mm thick. The 2.5 mm gap between the walls is used for active cooling with helium gas as the beam pipe is exposed to beam induced heating effects of a few hundred Watt. The outer cylinder is covered with a 20 μm thick gold foil to reduce background from synchrotron radiation.

The described configuration was used from 1999 to 2003 and replaced during the SVD upgrade process. The radii of the beam pipe used from 2003 to 2010 were reduced such that the inner radius is 15 mm.

A detailed description of the beam pipe is given in Ref. [4].

3.2.2 Silicon Vertex Detector

To achieve one of the design goals, the measurement of time-dependent *CP* violation in neutral B-meson decays, an excellent vertex resolution is required. The Lorentz boost of the KEKB accelerator increases the decay length difference of the two B mesons to about 200 μm. The silicon vertex detector (SVD) provides a spatial resolution for the z-vertex position of about 100 μm and allows for precision measurements.

The SVD configuration used from 1999 to 2003, referred to as SVD1, consists of three layers of double-sided silicon strip detectors (DSSD), see Fig. 3.4. The DSSDs are depleted pn-junctions and passing charged particles create electron hole pairs along their trajectory that drift to the n^+ and p^+ strips on the DSSD surface. The n^+ and p^+ strips are aligned perpendicular and parallel to the beam direction, respectively. In combination they provide a measurement of charged tracks in $r\phi$ and z direction. The innermost layer is mounted as close as possible to the beam pipe

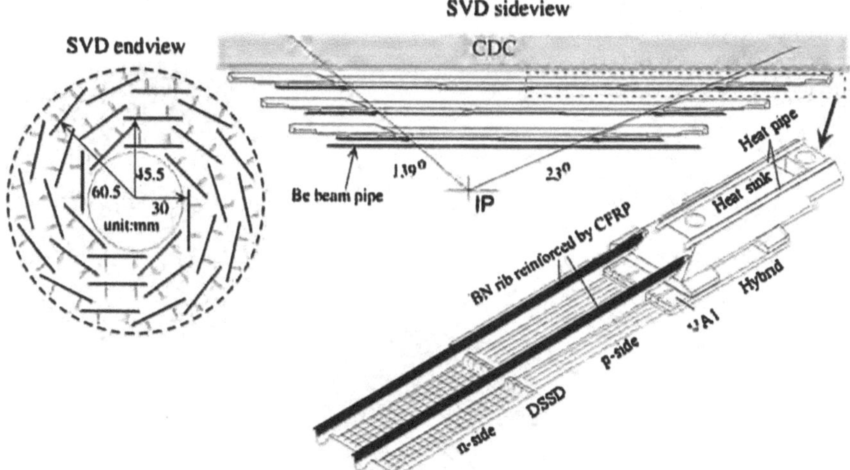

Fig. 3.4 Detector configuration of the SVD1 used from 1999 to 2003. Taken from Ref. [4]

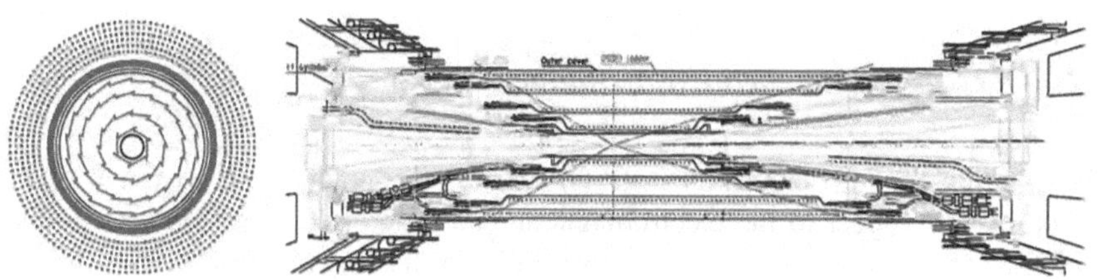

Fig. 3.5 Detector configuration of the SVD2 used from 2003 to 2010. Taken from Ref. [6]

with a radius of 30 mm, followed by layer two and three with a radius of 45.5 and 60.5 mm, respectively. The SVD1 covers a polar angle of $23° < \theta < 139°$.

The detector was upgraded in 2003 with an improved version, referred to as SVD2, consisting of four layers equipped with DSSDs, see Fig. 3.5. During the upgrade the beam pipe was replaced and its diameter reduced to decrease the distance from the interaction point to the first detector layer. The four layer radii of the SVD2 are 20, 43.5, 70 and 80 mm. The angular coverage was also improved and covers a polar angle of $17° < \theta < 150°$. The upgrade improved the z-vertex resolution for decays with low momentum tracks by about 20 % with respect to the SVD1.

A detailed description of the different silicon vertex detectors used in the Belle detector is given in Refs. [4, 6, 7].

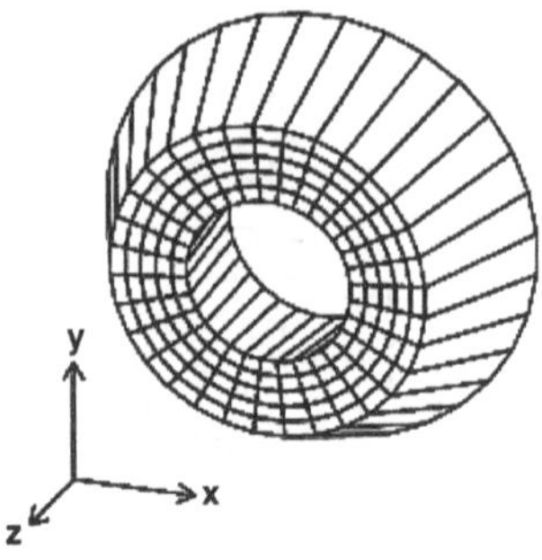

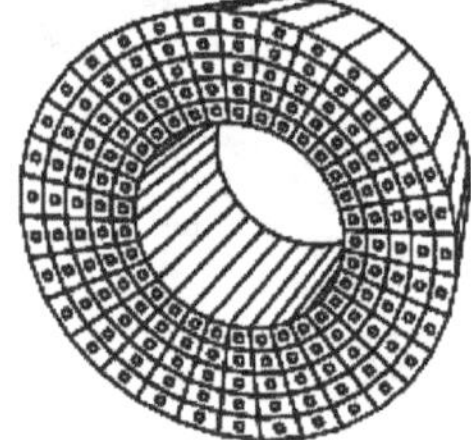

Fig. 3.6 An isometric view of the BGO crystals of the forward and backward EFC detectors. Taken from Ref. [4]

3.2.3 Extreme Forward Calorimeter

The extreme forward calorimeter (EFC) extends the polar angle coverage of the detector with respect to the electromagnetic calorimeter. The EFC covers a polar angle of $6.4° < \theta < 11.5°$ in the forward direction and $163.3° < \theta < 171.2°$ in the backward direction. It detects electrons and photons and further serves as beam monitor for the KEKB accelerator and luminosity measurement device for the Belle detector.

Due to its proximity to the interaction region and its location in the extreme forward and backward direction, it is exposed to very high levels of radiation. Bismuth germanium oxide (BGO) ($Bi_4Ge_3O_{12}$) is used for the crystals to fulfil the requirements on radiation hardness. The scintillation light is collected by photo diodes. The arrangement of the BGO crystals is shown in Fig. 3.6 and approximately such that each crystal points towards the interaction point.

A detailed description of the EFC is given in Ref. [4].

3.2.4 Central Drift Chamber

The central drift chamber (CDC) provides measurements of the trajectories of charged particles. The trajectories are bend in the 1.5 T magnetic field and from their curvature a momentum measurement of the reconstructed tracks is possible. Further, the measurement of specific energy loss of charged particles by ionisation, dE/dx, provides information that can be used for particle identification.

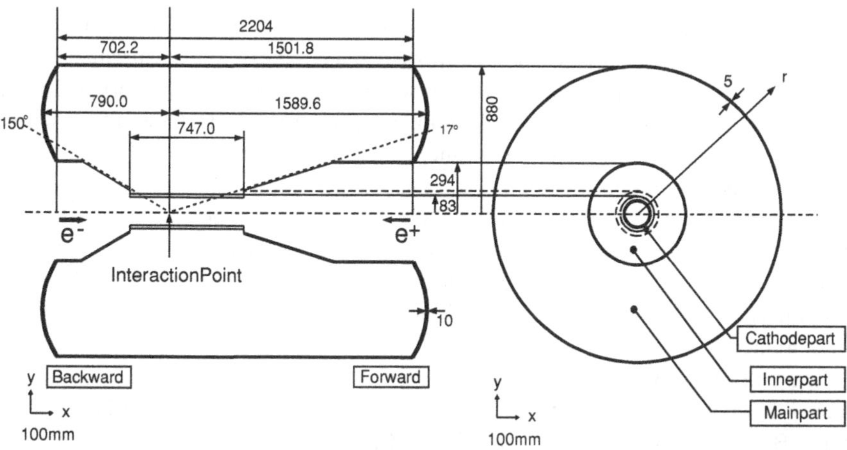

Fig. 3.7 Geometry of the CDC. The lengths in the figure are in units of mm. Taken from Ref. [4]

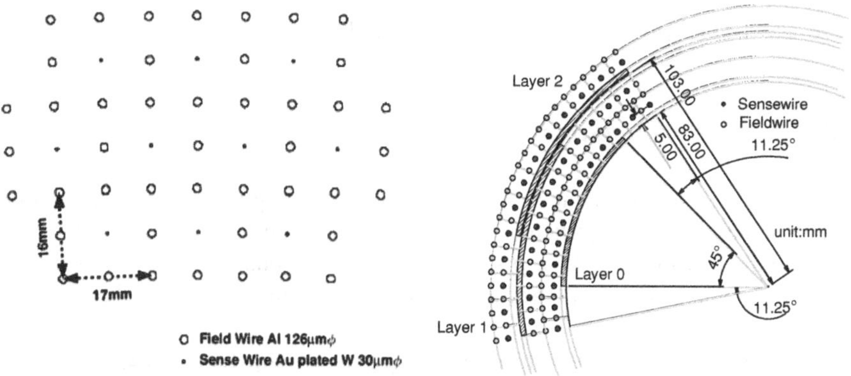

Fig. 3.8 Cell structure and arrangement of wires in the CDC. Taken from Ref. [4]

The CDC is constructed as cylindrical wire drift chamber with an asymmetry in z-direction to account for the Lorentz boost of the center-of-mass system of the collision. It is about 2.4 m long with an inner radius of 83 mm, an outer radius of 888 mm and covering a polar angle of $17° < \theta < 150°$. The geometry details are shown in Fig. 3.7. The CDC has 12 cylindrical superlayers, each containing between three and six axial or small-angle stereo layers, and three cathode strip layers. In total, the CDC has 8400 drift cells, each having 8 negatively biased field wires providing an electrical field that surrounds a positively biased sense wire. The cell structure and the arrangement are shown in Fig. 3.8.

The CDC is filled with a gas mixture of 50 % helium and 50 % ethane. The low-Z gas mixture minimizes the effect of multiple Coulomb scattering and has further the advantage of a small photo-electric cross-section, reducing the synchrotron radiation background. The large ethane fraction allows for a good dE/dx measurement.

Charged particles passing the CDC ionise the gas along their trajectory and electron and ions drift towards anode and cathode wires, respectively. Close to the wires, due to the strong electric field, an avalanche effect occurs that amplifies the electric pulse detected by the sense wires. The axial layers provide a position measurement in the $r\phi$-plane for measurement of the transverse momentum, whereas the small-angle stereo layers provide additional z-position information. Track finding algorithms described in Ref. [8] combine the position measurements to reconstructed tracks. The tracks can further be combined with hits in the SVD to improve the momentum measurement.

The amplitude of the electric pulse in the hit wires is also used to measure the energy loss of the charged particle. The dE/dx distribution is described by the Bethe-Bloch formula (see Ref. [9, Chap. 30]) that depends on the velocity of a particle. Consequently, the dE/dx measurement allows to distinguish among different charged particles as the velocity depends on the particle momentum and mass. The dE/dx measurement provides good separation between kaons and pions up to particle momenta of $p \approx 1.5\,\text{GeV}$.

A detailed description of the CDC is given in Refs. [4, 10].

3.2.5 Aerogel Čerenkov Counter

The aerogel Čerenkov Counter (ACC) is a detector for providing information on particle identification, in particular to separate kaons from pions. It is sensitive to a momentum range of $1.2\,\text{GeV} < p < 3.5\,\text{GeV}$ and thus complementary to the dE/dx measurement in the CDC and the time-of-flight measurement by the TOF, described in the Sect. 3.2.6.

Charged particles that pass a medium with a velocity larger than the speed of light in the medium c_{medium} radiate a cone of Čerenkov light, similar to the supersonic cone of a supersonic aircraft. The speed of light in the medium is related to the refractive index n of the medium by $c_{\text{medium}} = c_{\text{vacuum}}/n$. For a particle with mass m, momentum p and velocity β, Čerenkov light is emitted if

$$n > \frac{1}{\beta} = \sqrt{1 + \left(\frac{m}{p}\right)^2}. \tag{3.1}$$

The material of the ACC is chosen such that for momenta $p > 1.2\,\text{GeV}$ pions, electrons and muons emit Čerenkov light, whereas the velocity of heavier kaons and protons with the same momenta is below the threshold velocity. The ACC is a threshold counter and does not image the cone of the Čerenkov light, which would provide further information as it is related to the particles velocity.

The ACC threshold counters cover a polar angle of $17° < \theta < 127°$ and are installed in the barrel and forward endcap region, see Fig. 3.9. The refractive index

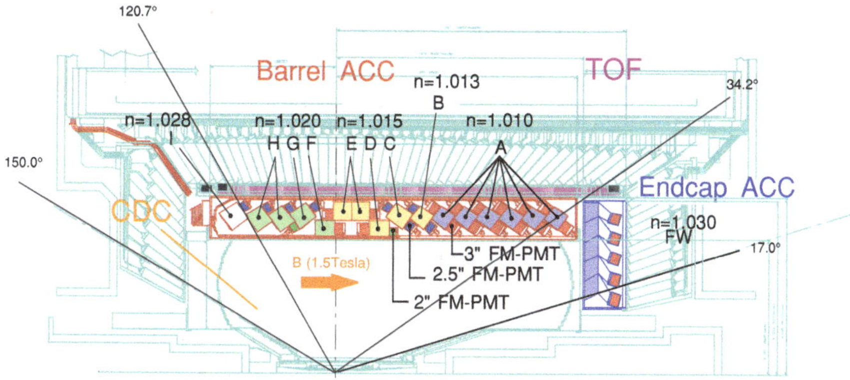

Fig. 3.9 Arrangement of the ACC in the Belle detector. Taken from Ref. [4]

of the silica aerogel is $1.010 \leq n \leq 1.030$ and depends on the polar angle. Each module is arranged such that it points toward the interaction region and the Čerenkov light is detected by fine-mesh photomultipliers.

A detailed description of the ACC is given in Refs. [4, 11].

3.2.6 Time-of-Flight Counter

The time-of-flight (TOF) detector system measures the time that a particle needs to travel from the interaction point to the TOF module. In practice, the time difference to the bunch crossing, known with high precision from the accelerator, is measured. In combination with a momentum measurement this allows to deduce the particle mass. The mass m of a particle is related to the measured time-of-flight T by

$$m = \frac{p}{c}\sqrt{\left(\frac{cT}{L}\right)^2 - 1}, \tag{3.2}$$

where p is the momentum of the particle, measured in the CDC and SVD, and L the helical distance travelled by the particle from the interaction point to the TOF module.

The TOF system is sensitive to momenta $p < 1.2\,\text{GeV}$ and complementary to the ACC as it provides information on the separation of kaons from pions in lower momentum ranges. Due to its good time resolution of about 100 ps the TOF is also used as timing signal for the trigger system.

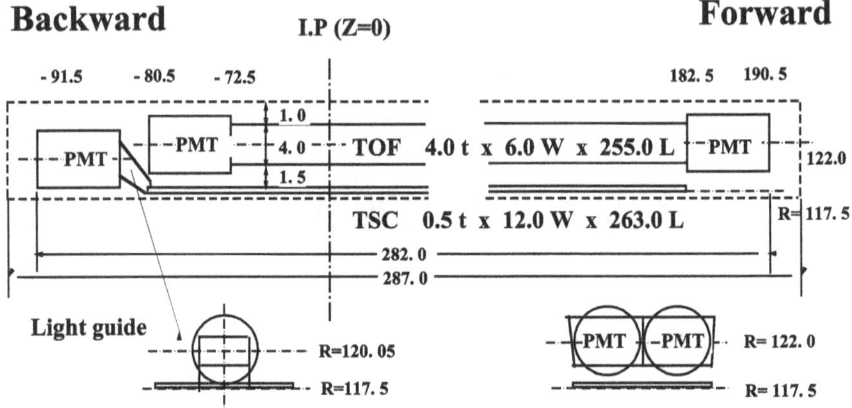

Fig. 3.10 Layout of a TOF detector module. Taken from Ref. [4]

The TOF modules are plastic scintillation counters with attached photomultiplier tubes, see Fig. 3.10. The modules are installed in the barrel region in a radial distance of 1.2 m with respect to the interaction point and cover a polar angle of $34° < \theta < 120°$.

A detailed description of the TOF is given in Refs. [4, 12].

3.2.7 Electromagnetic Calorimeter

The electromagnet calorimeter (ECL) measures the energy and position of electrons and photons. In electromagnetic showers, cascades of bremsstrahlung and pair production processes, the energy of the particles is absorbed in the ECL. The ECL is made of 8376 scintillating cesium iodide crystals doped with thallium, CsI(Tl), as wavelength shifter and read out with photodiodes. The crystals provide a good energy resolution, typically $(1 - 2)$ %, from a few dozen MeV up to about 4 GeV, thus covering the full energy range relevant for the Belle experiment. The large number of crystals provides a good position resolution and covers a polar angle of $12° < \theta < 155°$, except a small gap between the barrel and endcap regions for construction reasons. The geometry of the ECL and the arrangement of the ECL crystals is shown in Fig. 3.11.

The measurement in the ECL is also used to contribute to the particle identification. The ratio of deposited energy to momentum of a charged track is close to unity for electrons, whereas it is smaller for others. Further, the shower shape of hadrons differs from those of electrons.

A detailed description of the ECL is given in Refs. [4, 13].

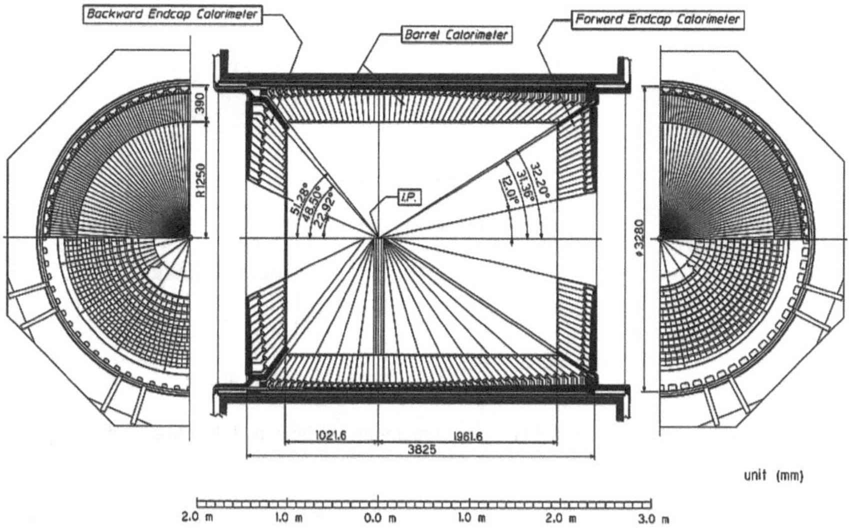

Fig. 3.11 Geometry and arrangement of the CsI(Tl) crystals in the ECL. Taken from Ref. [4]

3.2.8 Superconducting Solenoid Magnet

The superconducting solenoid magnet has a cylindrical volume of 3.4 m in diameter and 4.4 m in length. It provides a 1.5 T magnetic field and covers all detector components except the KLM, which is located outside in the iron support structure. The iron support structure also serves as a return path for the magnetic flux.

The superconducting coil is made of a niobium-titanium-copper (NbTi/Cu) alloy, that is stabilized by aluminium. The cryostat uses liquid helium to achieve superconducting temperatures. The magnet has a nominal current of 4400 A and stores an energy of 35 MJ. The layout of the magnet and a cross-section of the coil is shown in Fig. 3.12.

A detailed description of the superconducting solenoid is given in Ref. [4].

3.2.9 K^0_L *and Muon Detector*

The K^0_L and muon detector (KLM) is located in the iron support structure of the Belle detector. It is designed to identify muons and long-lived neutral kaons. The KLM is composed of alternating layers of 4.7 cm thick iron plates from the support structure and superlayers of resistive plate counters (RPC) that detect the charged particles. The barrel (endcap) contains 14 iron layers and 15 (14) RPC superlayers, and covers a polar angle of $20° < \theta < 155°$.

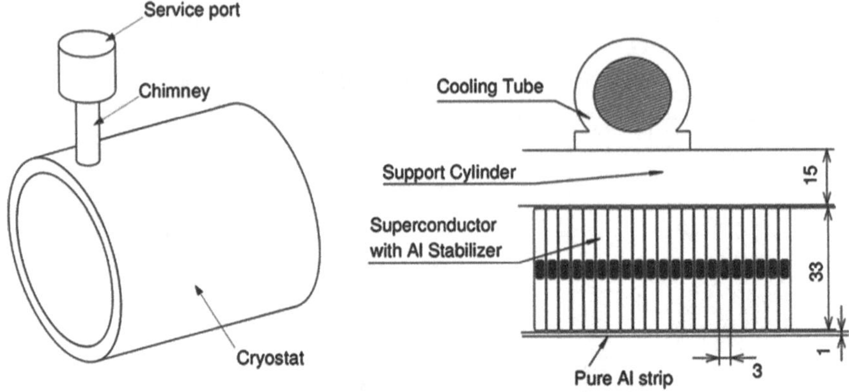

Fig. 3.12 Layout of the Magnet and cross-sectional view of the coil. Taken from Ref. [4]

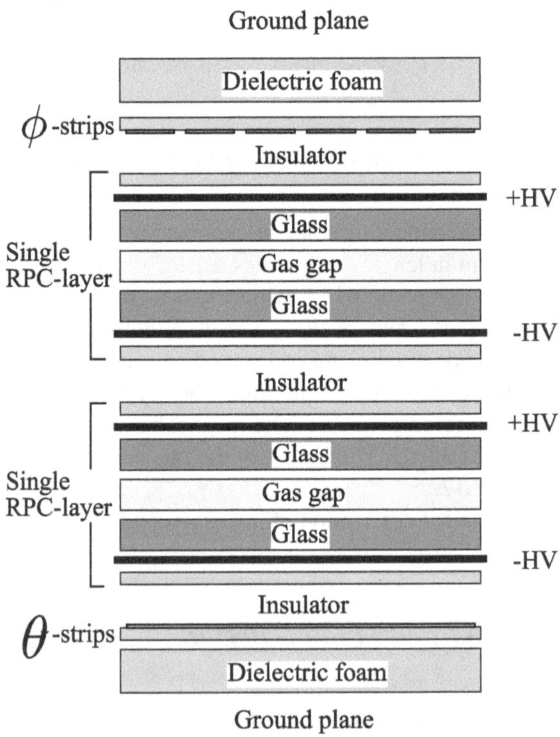

Fig. 3.13 Cross section of a RPC superlayer in the KLM detector. Taken from Ref. [4]

The layout of a RPC superlayer is shown in Fig. 3.13. Each superlayer has two RPC modules, that consist of two glass-electrodes with a high bulk resistivity (10^{10} Ω cm) and a gas filled gap. A charged particle passing the gap initiates a streamer in the gas that creates a local discharge of the plates. The discharge induces a signal on the external read out strips in ϕ and θ direction that are used to record location and time of the ionization.

The iron absorbers in the KLM provide 3.7 interaction lengths, with an additional 0.8 interaction lengths from the ECL, of material to convert K_L^0 mesons into showers of ionising particles. The KLM measures only the shower direction and thus the K_L^0 flight direction. As it provides no energy measurement the KLM mainly is used as a veto system. Showers that can not be matched to an extrapolated track from the CDC are likely due to long-lived neutral hadrons.

The KLM is also used for particle identification of weakly interacting muons. With sufficient momentum, muons pass all other detector components, only deflected by multiple scattering, and penetrate several layers of the KLM. Reconstructed tracks from the CDC that can be matched to a series of hits in the KLM are very likely from muons.

A detailed description of the KLM is given in Refs. [4, 14].

3.2.10 Trigger and Data Acquisition System

The data acquisition (DAQ) and storage of the data from the Belle detector is controlled by the trigger system. The trigger system has to decide on the basis of fast signals from the detector components whether an event is kept and recorded or discarded. Events of interest are hadronic $\Upsilon(4S)$ decays, $e^+e^- \to q\bar{q}$ ($q \in \{u, d, s, c\}$) continuum events, two photon processes, $e^+e^- \to \tau^+\tau^-$, Bhabha scattering, and others. Discarded are events from synchrotron radiation, interactions from the beam with residual gas in the beam pipe or events caused by cosmic rays.

At a peak luminosity of $\mathcal{L} = 2.1 \times 10^{34}\,\mathrm{cm}^{-2}\,\mathrm{s}^{-1}$ the summed rate of signal (background) processes is about 200 Hz (600 Hz). The trigger system is a multi-level system. In the first layer information from the subdetectors is collected online and it is decided whether an event is passed to the next level, that runs fast track finding algorithms. If reconstructed tracks originate from the interaction region, an event is kept. The efficiency for hadronic events is more than 99 %, while the data rate is reduced by a factor two at this level. The final level processes the data offline and performs the full events reconstruction and particle identification. Further, a set of minimal selection criteria is applied before the data is stored for analysis.

A detailed description of the trigger and DAQ system is given in Refs. [4, 15–17].

3.2.11 Illustration of a Reconstructed Event

To illustrate the interplay of the individual detector components a reconstructed event is shown in Figs. 3.14 and 3.15 from two different perspectives.

The event contains a reconstructed $B^0 \to \phi K^*$ candidate that is associated with a high signal probability. The individually reconstructed tracks of the decay chain $B^0 \to \phi K^* \to (K^+K^-)(K^+\pi^-)$ are highlighted in orange, whereas the black tracks

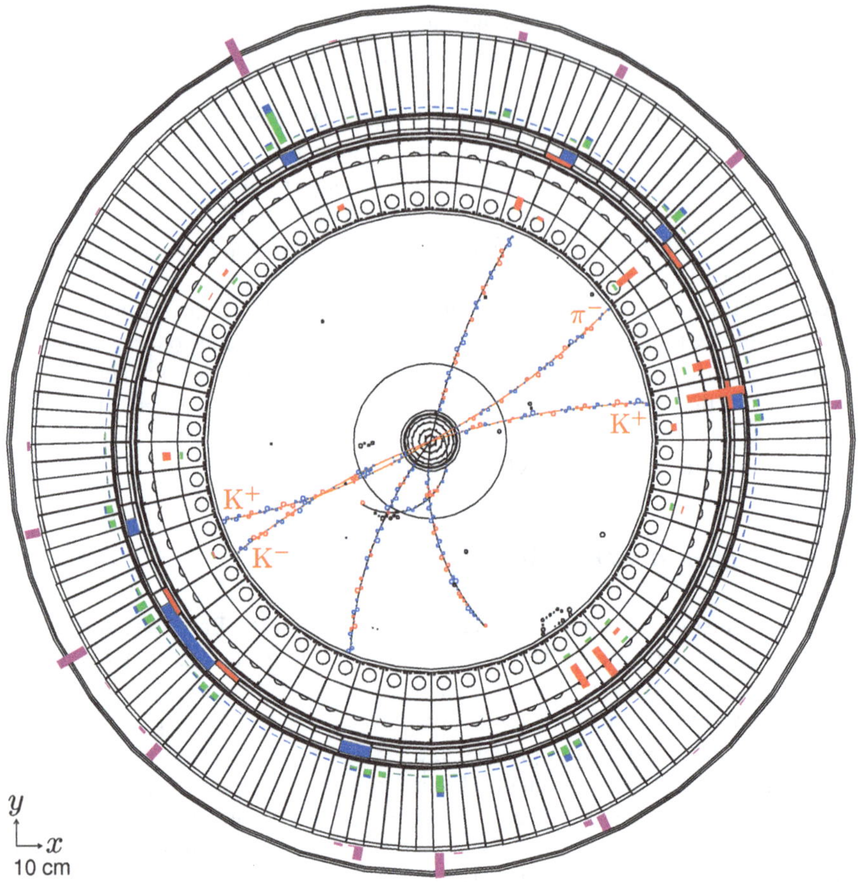

Fig. 3.14 Projection of a reconstructed event, recorded on 23rd March 2005, on the xy-plane with a signal candidate (*orange tracks*) that is associated with a high signal probability

originate from the other B meson in the event. Measured drift times in the CDC are illustrated by red and blue circles along the tracks, whereas the detector response in the ACC, TOF and ECL is illustrated by differently colored bars. Only detector components inside the solenoid are shown.

3.3 Data Samples

Data

This analysis uses the full data sample of 711 fb^{-1} recorded at the $\Upsilon(4S)$ resonance. The data sample corresponds to $(772 \pm 11) \times 10^6$ $\text{B}\overline{\text{B}}$ pairs, referred to as $\text{B}\overline{\text{B}}$ events. The remaining events in the data sample are from $\text{e}^+\text{e}^- \rightarrow \text{q}\bar{\text{q}}$ ($\text{q} \in \{\text{u}, \text{d}, \text{s}, \text{c}\}$) events, refereed to as continuum events.

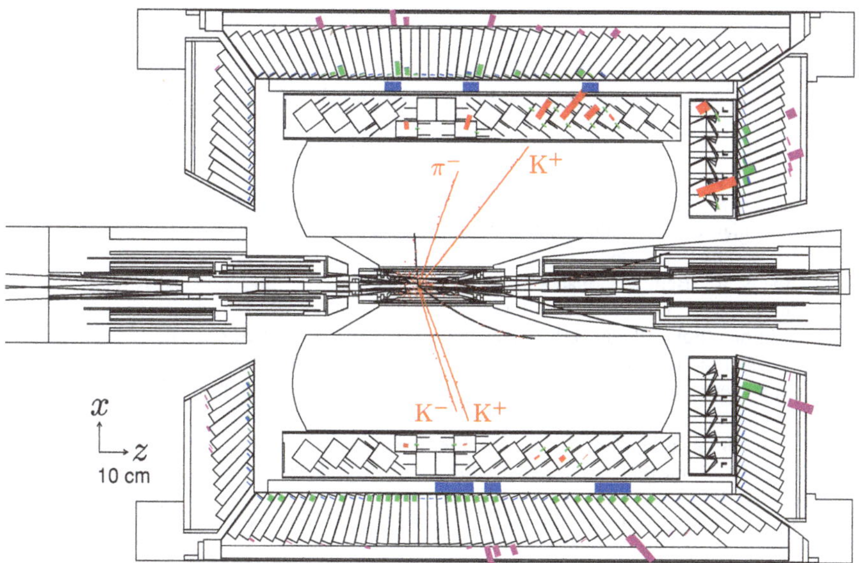

Fig. 3.15 Projection of a reconstructed event, recorded on 23rd March 2005, on the xz-plane with a signal candidate (*orange tracks*) that is associated with a high signal probability

Further, this analysis uses the data sample of 79 fb^{-1} recorded below the $\Upsilon(4S)$ resonance, referred to as off-resonance data. The center-of-mass energy of these events is below the $\mathrm{B\overline{B}}$ threshold, thus only $\mathrm{e^+e^-} \rightarrow \mathrm{q\bar{q}}$ ($\mathrm{q} \in \{\mathrm{u, d, s, c}\}$) events contribute.

Simulated Data

The analysis procedure was established using Monte-Carlo (MC) simulated events. The program packages EvtGen [18] and Pythia [19] are used to simulate the decay processes, while the detector response is simulated using GEANT3 [20]. The PHOTOS package [21] is used to take into account final state radiation. The simulated decays are generated according to known branching fractions and all cross sections are scaled according to the integrated luminosity recorded by Belle.

The statistics of continuum $\mathrm{e^+e^-} \rightarrow \mathrm{q\bar{q}}$ ($\mathrm{q} \in \{\mathrm{u, d, s, c}\}$) and inclusive $\mathrm{b} \rightarrow \mathrm{c}$ events corresponds to four times the data statistics. These samples are referred to as the four streams of continuum and $\mathrm{B\overline{B}}$ MC. In addition, a MC sample of inclusive $\mathrm{b} \rightarrow \mathrm{s}$ decays with 50 times the statistics in data is used and referred to as rare MC.

Signal events are either simulated as three-body phase space decay $\mathrm{B^0} \rightarrow \phi\mathrm{K^+}\pi^-$ or via the intermediate states $\mathrm{B^0} \rightarrow \phi(\mathrm{K}\pi)^*_0$, $\mathrm{B^0} \rightarrow \phi\mathrm{K^*(892)^0}$, and $\mathrm{B^0} \rightarrow \phi\mathrm{K}^*_2(1430)^0$ for S-, P-, and D-wave, respectively. In total, several million simulated signal events are used and have been produced for this analysis. The different samples are generated with various polarizations, each corresponding to multiple times the statistics in the data sample. The samples are e.g. used to implement and test the event reconstruction and study reconstruction efficiencies.

References

1. S. Kurokawa, E. Kikutani, Overview of the KEKB Accelerators. Nucl. Instr. Meth. A **499**(1), 1–7 (2003)
2. KEKB B-Factory Design Report, KEK report 95-7. High Energy Accelerator Research Organization, Japan (1995)
3. K. Akai et al., Commissioning of KEKB. Nucl. Instr. Meth. A **499**(1), 192–227 (2003)
4. A. Abashian et al., (Belle Collaboration), The Belle detector. Nucl. Instr. Meth. A **479**(1), 117–232 (2002)
5. T. Abe et al., (Belle II Collaboration), Belle II technical design report. KEK report 2010-1. High Energy Accelerator Research Organization, Japan (2010)
6. Y. Ushiroda, Belle silicon vertex detectors. Nucl. Instr. Meth. A **511**(1–2), 6–10 (2003) (Proceedings of the 11th International Workshop on Vertex Detectors)
7. Z. Natkaniec et al., Status of the Belle silicon vertex detector. Nucl. Instr. Meth. A **1**(560), 1–4 (2006)
8. Belle Tracking Group, Charged particle tracking at Belle. Belle note, #327 (2000)
9. J. Beringer et al., (Particle Data Group), The review of particle physics. Phys. Rev. D **86**, 010001 (2012), http://pdg.lbl.gov
10. H. Hirano et al., A high-resolution cylindrical drift chamber for the KEK B-factory. Nucl. Instr. Meth. A **455**(2), 294–304 (2000)
11. T. Iijima et al., Aerogel Cherenkov counter for the Belle detector. Nucl. Instr. Meth. A **453**(1), 321–325 (2000)
12. H. Kichimi et al., The Belle TOF system. Nucl. Instr. Meth. A **494**(1), 298–302 (2002)
13. K. Miyabayashi, Belle electromagnetic calorimeter. Nucl. Instr. Meth. A **494**(1), 298–302 (2002)
14. A. Abashian et al., Muon identification in the Bell experiment at KEKB. Nucl. Instr. Meth. A **491**(1), 69–82 (2002)
15. Y. Ushiroda et al., Development of the central trigger system for the Belle detector at the KEK B-factory. Nucl. Instr. Meth. A **438**(2), 460–471 (1999)
16. S.Y. Suzuki et al., The Belle DAQ system. Nucl. Instr. Meth. A **453**(1), 440–444 (2000)
17. S.Y. Suzuki et al., Belle DAQ system upgrade at 2001. Nucl. Instr. Meth. A **494**(1), 535–540 (2002)
18. D.J. Lange, The EvtGen particle decay simulation package. Nucl. Instr. Meth. A **462**(1–2), 152–155 (2001)
19. T. Sjstrand et al., High energy physics event generation with PYTHIA 6.1. Comput. Phys. Commun. **135**, 238–259 (2001)
20. R. Brun et al., GEANT3 User's Guide. CERN, Report, DD/EE/84-1 (1987)
21. E. Barberio, Z. Was, PHOTOS: a universal Monte Carlo for QED radiative corrections. Version 2.0. Comput. Phys. Commun. **79**, 291–308 (1994)

Chapter 4
Analysis Methods and Tools

In this chapter a brief introduction into the concept of the maximum likelihood (ML) method is given, which is used to obtain the physical results presented in this thesis. Two problems in the context of ML, namely calculation of normalization integrals and dependencies among observables, and developed solutions are discussed. Further, an alternative approach for error propagation is presented. Finally, details on the continuum suppression, a technique to address the dominant background in the analysis presented in this thesis, are given.

4.1 Maximum Likelihood Method

A common method to extract an estimator for a set of parameters of interest $\vec{\mu}$ from an observed distribution of events is the ML method or fit. Given a model $f(\vec{x};\vec{\mu})$, with $\vec{x}$ being a set of observables in the data, the parameters $\vec{\mu}$ are optimized such that the model becomes most likely for the observed data distribution. The model could e. g. be a distribution predicted by theory or an empirically determined distribution from simulated or real data.

Usually, the model is chosen to be a probability density function (PDF) $\mathcal{P}(\vec{x};\vec{\mu})$, which is defined to be positive and normalized to unity over the allowed range of $\vec{x}$ for all possible parameters $\vec{\mu}$, i. e.

$$\forall\vec{\mu} : \int \mathcal{P}(\vec{x};\vec{\mu})d\vec{x} \equiv 1. \tag{4.1}$$

Models can involve an additional set of nuisance parameters $\vec{\vartheta}$ beside the parameters of interest. For a model $\mathcal{P}(\vec{x};\vec{\mu};\vec{\vartheta})$ derived from theory such nuisance parameters could e. g. be the detector resolution or external parameters from other measurements.

The likelihood $L(\vec{\mu};\vec{\vartheta})$ is defined as the PDF evaluated at a measured data point $\vec{x}_j$:

M. Prim, *Polarization and CP Violation Measurements*, Springer Theses,
DOI: 10.1007/978-3-319-05756-9_4,

$$L(\vec{\mu};\vec{\vartheta}) = \mathcal{P}(\vec{x}_j;\vec{\mu};\vec{\vartheta}). \tag{4.2}$$

For an ensemble of N independent data points, the likelihood of the ensemble is the product of the individual likelihoods for each measured data point:

$$L(\vec{\mu};\vec{\vartheta}) = \prod_{j=1}^{N} \mathcal{P}(\vec{x}_j;\vec{\mu};\vec{\vartheta}). \tag{4.3}$$

In practice, the logarithm of the likelihood function $\mathcal{L}(\vec{\mu};\vec{\vartheta})$ is used for numerical reasons and the minimum of the negative log-likelihood function is searched to obtain the most likely estimator for the parameters of interest $\vec{\mu}$. A data distribution might also consist of several components, such as signal and background; each with its own model. Furthermore, the number of expected events in an ensemble can be a parameter of interest which requires an extension of the likelihood function. Altogether this can be taken into account by the log-likelihood function of an unbinned extended ML fit, as it is used in this analysis and given by

$$\mathcal{L}(\vec{\mu};\vec{\vartheta}) = \ln L(\vec{\mu};\vec{\vartheta}) = \sum_{j=1}^{N} \ln \left\{ \sum_{i=1}^{N_c} N_i \mathcal{P}_i(\vec{x}_j;\vec{\mu};\vec{\vartheta}) \right\} - \sum_{i=1}^{N_c} N_i, \tag{4.4}$$

where N_c is the number of different components in the data set, N_i is the expected number of events for the ith component, and $\mathcal{P}_i$ is the PDF for the ith component.

The main problem of the ML method is the choice of the correct model. If e. g. a quadratic distribution $f(x) = x^2$ is fitted with a linear model $f(x) = a \cdot x$, the ML estimator of the slope parameter a will not provide any useful information nor will the model be able to describe the data distribution. If no model can be derived from theory, it is a common practice to derive a model from the data distribution itself; either by using non-parametric histograms from, e. g., simulated data or by determining a parametric description.

In this analysis the negative log-likelihood function is minimized using the MINUIT [1] algorithm in the RooFit [2] package of the ROOT framework [3]. The RooFit package provides a large set of PDFs that can be used to build a model for experimentally observed distributions and can be extended by analysis specific PDFs. Furthermore, it provides functionality to normalize PDFs and visualize fit results.

The ML estimators are ideal estimators in the limit of infinite statistics. They are consistent, meaning that they give the correct value in the limit of infinite statistics, and efficient, meaning that the variance of the estimated value is minimal.

A derivation of the ML method from basic principles and the properties of a ML estimator can be found in Ref. [4] or most textbooks on statistics and data analysis. A review on the ML method and basic concepts of probability and statistics can be found in Ref. [5, Chaps. 35 and 36].

4.1.1 Multidimensional Maximum Likelihood Fit

In the case, that the dimension of the set of observables $\vec{x}$ in Eq. (4.4) is greater than one, the ML fit becomes a multidimensional ML fit. The ML method has no constraints on the dimension of $\vec{x}$ as long as a corresponding multidimensional model is given. Again, theory might partially provide a model for signal distributions (as it is shown in Sect. 2.5.3), but often distributions need to be derived from data.

Given a three-dimensional set of observables $\vec{x} = \{x, y, z\}$, a common method to describe the three-dimensional PDF is a product of marginal distributions

$$\mathcal{P}(\vec{x}; \vec{\mu}) = \mathcal{P}(x; \vec{\mu}_x)\mathcal{P}(y; \vec{\mu}_y)\mathcal{P}(z; \vec{\mu}_z), \tag{4.5}$$

where $\mathcal{P}(i, \vec{\mu}_i)$ is the marginal distribution of observable i in $\vec{x}$ and $\vec{\mu}_i$ are the parameters sensitive to this observable. This procedure is valid for independent observables. Indeed, Eq. (4.5) is the definition of independence among observables.

If the observables are not independent a more sophisticated model is required (see e. g. Sect. 6.2). Using conditional PDFs is one method to construct multidimensional models. Assuming a dependence between x and y the model can be written as

$$\mathcal{P}(\vec{x}; \vec{\mu}) = \mathcal{P}(x|y; \vec{\mu}_x)\mathcal{P}(y; \vec{\mu}_y)\mathcal{P}(z; \vec{\mu}_z), \tag{4.6}$$

with the conditional PDF $\mathcal{P}(x|y; \vec{\mu}_x)$ being normalized such that

$$\forall y, \vec{\mu}_x : \int \mathcal{P}(x|y; \vec{\mu}_x)dx \equiv 1. \tag{4.7}$$

For an n-dimensional set of observables $\vec{x}$, there exist $(n^2 - n)/2$ possible dependencies among the observables. Constructing a parametric conditional PDF can become a complex task even for $n \geq 2$. Using non-parametric multidimensional histograms requires exponentially increasing amounts of statistics with increasing n. Assuming m bins per dimension, the multidimensional histogram would have m^n bins in total. To reduce the uncertainties on such a non-parametric distribution, as it can be derived e. g. from simulated data, each bin has to have sufficient statistics which typically requires a multiple of m^n events to be simulated. Again, this can easily become a complex problem for $n \geq 2$.

4.2 Optimization of Numeric Integration

The normalization integrals in Eq. (2.27) require a four-dimensional numeric integration, which is computationally intensive, to normalize the PDF. As the weights in $\mathcal{M}$ are adjusted during an ML fit, this operation needs to be performed thousands of times. Such integrations can however be optimized dramatically when certain conditions are satisfied.

The integration over a simple matrix element $|\mathcal{M}|^2$ with two amplitudes $\mathcal{A}_i(\vec{x})$ ($i = 0, 1$) depending on observables $\vec{x}$ and their complex weights $A_i = a_i e^{i\varphi_i}$

$$\int |\mathcal{M}|^2 d\vec{x} = \int |A_0 \cdot \mathcal{A}_0(\vec{x}) + A_1 \cdot \mathcal{A}_1(\vec{x})|^2 d\vec{x}, \tag{4.8}$$

can be expanded to

$$\begin{aligned} \int |\mathcal{M}|^2 d\vec{x} =& a_0^2 \int |\mathcal{A}_0(\vec{x})|^2 d\vec{x} + a_1^2 \int |\mathcal{A}_1(\vec{x})|^2 d\vec{x} \\ &+ 2a_0 a_1 \cos \Delta\varphi \int \mathrm{Re}\{\mathcal{A}_0(\vec{x})\mathcal{A}_1^*(\vec{x})\} d\vec{x} \\ &- 2a_0 a_1 \sin \Delta\varphi \int \mathrm{Im}\{\mathcal{A}_0(\vec{x})\mathcal{A}_1^*(\vec{x})\} d\vec{x}, \end{aligned} \tag{4.9}$$

with $\Delta\varphi = \varphi_0 - \varphi_1$. Given n amplitudes $\mathcal{A}_i$, one always obtains n integrals over $\mathcal{A}_i$ squared and $(n^2 - n)/2$ integrals over the real and imaginary parts of the product of two amplitudes, respectively. If the amplitudes $\mathcal{A}_i$ have no free parameters, all integrals become constant as only the weights are adjusted.

In the context of the signal model, derived in Sect. 2.5.3, there are $n = 7$ helicity amplitudes, resulting in 49 constant integrals, as parameters like resonance masses, interaction radii and other similar quantities are fixed. These integrals are computed once with high precision and then are used on demand, thereby significantly reducing the amount of CPU time. This method is several orders of magnitude faster than a numeric integration in each iteration of the fit.

For cross-checks, a comparison between this optimized approach and the numeric integration was performed. This exercise confirmed the validity of the approach but required several days of CPU time.

The same technique can also be used to improve the computation of projection integrals onto one dimension d in $\vec{x}$ for a fixed value of x_d. In a typical projection plot, hundreds of $(\dim \vec{x} - 1)$-dimensional integrations are necessary per plot, normally requiring several hours of CPU time. These integrals can be computed in parallel on a large scale cluster and stored. If loaded on demand, the improvement in speed is again several orders of magnitude.

A thread-safe implementation of this method has been developed that integrates into the RooFit [2] package but is not limited to it. It has become part of a software collection that is available within the Belle collaboration and has become part of the Belle II software stack. The user is required to implement the integrand, whereas on demand computation, serialization and deserialization of the results, and other technical issues are treated automatically.

4.3 Quantifying Dependence in Multivariate Data Sets

In a multidimensional ML fit it is crucial to use conditional PDFs where necessary and the product of marginal distributions where possible. For this analysis a method was developed to quantify the dependence among observables in a multivariate data set, thus providing guidance during the model building process. The content of this section has also been published in Ref. [6].

4.3.1 Linear Correlation Coefficient

The linear correlation coefficient r (see e. g. Ref. [4]), also known as the Pearson's correlation coefficient, can be used to describe the linear dependence between two observables x and y. For a given sample of N events, it can be computed from the data by

$$r = \frac{\sum_{i=1}^{N}(x_i - \bar{x})(y_i - \bar{y})}{\sqrt{\sum_{i=1}^{N}(x_i - \bar{x})^2}\sqrt{\sum_{i=1}^{N}(y_i - \bar{y})^2}}, \tag{4.10}$$

where

$$\bar{x} = \frac{1}{N}\sum_{i=1}^{N} x_i \quad \text{and} \quad \bar{y} = \frac{1}{N}\sum_{i=1}^{N} y_i \tag{4.11}$$

correspond to the sample mean in x and y. By construction, the possible values of r are within the interval $[-1, 1]$, where $r = 1(-1)$ corresponds to 100 % (anti-)linear dependence and $r = 0$ corresponds to no linear dependence.

In Fig. 4.1a a distribution of two observables generated randomly from a Gaussian normal distribution with no linear dependence is shown, whereas in Fig. 4.1b a distribution with linear dependence is shown. It is not possible to conclude from the absence of linear dependence that two observables are independent. In Fig. 4.1c a distribution is shown where x and y follow a circular distribution, i. e. $x = R \cdot \cos\phi$ and $y = R \cdot \sin\phi$. The linear correlation coefficient computed for this sample is zero.

This limitation of the linear correlation coefficient should be kept in mind when testing whether two observables are independent. Analyses involving observables from angular distributions can have distributions similar to the circular example with small correlation coefficients. Those distributions are not necessarily independent.

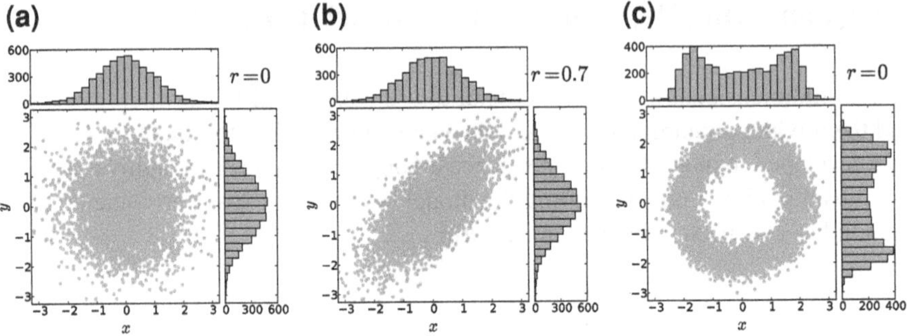

Fig. 4.1 Distributions of two observables x and y, randomly generated from a Gaussian normal distribution with **a** no and **b** 70 % linear dependence. In **c** an example of x and y being circular distributed is shown. The marginal distributions $\mathcal{P}(x)$ and $\mathcal{P}(y)$ of the observables x and y are shown *above* and *right* of the scatter plot, respectively

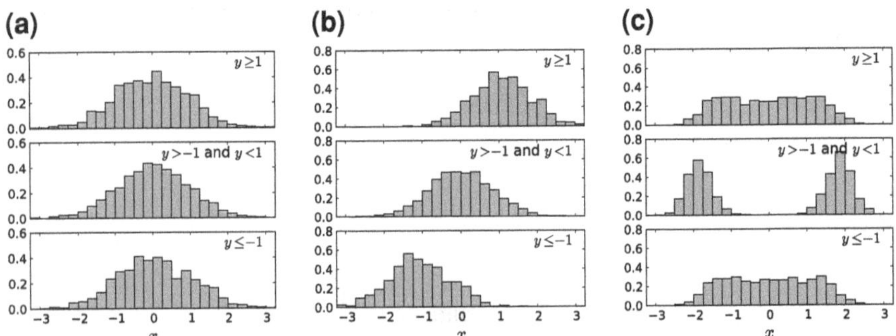

Fig. 4.2 Normalized projections on observable x in three different subranges of observable y for the three data sets **a, b** and **c** shown in Fig. 4.1

4.3.2 Projections in Subranges

A common practice to understand dependencies between two observables is the investigation of projections of one observable in subranges of the other. In Fig. 4.2 three examples of this method are given, using the same data sets as in Fig. 4.1.

In case of independent observables, the three projections show similar distributions. This method however does not allow to conclude independence in general. One has to be aware of symmetry axes in the original distribution. By choosing only two subranges with $y > 0$ and $y < 0$ instead of three, Fig. 4.2c would lead to two similar distributions. Using an adequate number of subranges avoids this problem in practical applications.

Although this method reveals strong dependencies, it requires further statistical tests such as for example a Kolmogorov-Smirnov-Test (see Ref. [4]) to judge whether two similar distributions are statistically compatible or not. Each pair of observables, with projections of one in n subranges of the other, requires $(n^2 - n)/2$ comparisons.

Given m observables and the $(m^2 - m)/2$ possible dependencies among them, the total number of comparisons becomes large and requires a reliable automatic procedure. Sorting different dependencies among different observables according to their importance is also not a simple task.

4.3.3 Hypothesis Test for Independence

What is actually desired to solve the original problem, namely to make a reliable statement on whether two distributions are independent or not, is a robust hypothesis test for independence. To describe such a test, the concept of Copulas, introduced in 1959 by Sklar for description on how a joint distribution function couples to its margins, is used. Sklar's theorem states:

Let S be a joint distribution function with margins F and G. Then there exists a copula C such that for all x,y in $\mathbb{R}$,

$$S(x, y) = C(F(x), G(y)). \tag{4.12}$$

If F and G are continuous, then C is unique; otherwise, C is uniquely determined on $Ran F \times Ran G$. Conversely, if C is a copula and F and G are distribution functions, then the function S defined by Eq. (4.12) *is a joint distribution function with margins F and G.*

Sklar's theorem and more details on copulas can be found for example in Ref. [7]. A special copula is the unit copula $C(u, v) = u \times v$, which connects the marginal distributions of independent observables, as can be seen by comparison to Eq. (4.5).

An algorithm to perform a robust hypothesis test for two observables x and y being independent in a given data set with N events is given by the following steps:

1. Determine the probability integral transforms $u = F(x)$ and $v = G(y)$ of observables x and y. First sort the data in x and y. The values of $u = I/N$ $(v = J/N)$, where I (J) is the index of x (y) in the sorted range, are then within the interval $[0, 1]$. This is sometimes referred to as flattening the distribution.
2. Create a $n \times n$ histogram $H(u, v)$ with bins of equal size and fill it with all events. The number of bins n should be chosen such that N/n^2 is large enough ($\gtrsim 25$). $H(u, v)$ corresponds to the empirical copula density.
3. In each bin of $H(u, v)$, if x and y are independent, $e = N/n^2$ entries are expected and the statistical uncertainty can be approximated by $\sigma_e = \sqrt{N/n^2}$ if the binning was chosen as suggested in step 2.
4. Compute the $\chi^2 = \sum_{i=1}^{n} \sum_{j=1}^{n} \frac{(h_{i,j} - e)^2}{\sigma_e^2}$, where $h_{i,j}$ is the content of the (i, j)th bin of $H(u, v)$.
5. The probability of the data being consistent with a flat hypothesis and thus x and y being independent observables follows a χ^2 distribution with $n^2 - (2n - 1)$ degrees of freedom. By construction, the number of degrees of freedom is reduced by $(2n - 1)$ due to the flatness of the two marginal distributions.

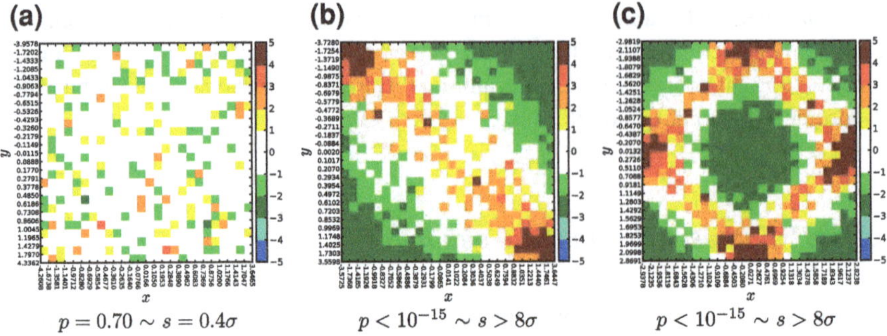

Fig. 4.3 Deviation in units of σ_e for the histogram $H(u, v)$ from a flat distribution for the three data sets shown in Fig. 4.1. The axis labels correspond to the untransformed (original) values of x and y, which allow for a simpler interpretation than the values in u and v. The resulting probability p for the distribution being consistent with a flat distribution, thus independent, and its transformation into significance s in units of standard deviations for being dependent is given below each figure

In short, the algorithm performs a test of $H(u, v)$ being consistent with the constant density $c(u, v) = \frac{\partial^2 C(u,v)}{\partial u \partial v}$ expected from the unit copula. The algorithm is able to identify any linear or non-linear dependence. The probability of the hypothesis can easily be compared among different pairs of observables in a multivariate data set with more than two observables. It also can be translated into the unit of standard deviations significance for the hypothesis that x and y are dependent (see the section about significance tests in Ref. [5, Sect. 36.2.2]). Examples of the resulting deviations from a flat distribution for the histogram $H(u, v)$ are shown in Fig. 4.3 for the data sets shown in Fig. 4.1.

The algorithm is very robust and delivers reliable results as it is based on rank statistics, no matter whether observable values are located on a small interval or reach over several orders of magnitude.

Another feature of this algorithm is the fact that its output scales with the size of the data set. A dependence might be negligible for low statistics but significant for higher statistics. Imagine for example a chessboard like distribution. Neither the algorithm nor the maximum likelihood fit will be sensitive to this dependence with low statistics and a simple product of marginal distributions will describe the data. With increasing statistics this dependence will become more and more significant as the size of the bins decreases. Also the fit model will have to be adjusted once the dependence reaches a certain level.

4.3.4 CAT: A Correlation Analysis Toolkit

The methods described in this section are implemented in a correlation analysis toolkit (CAT) that is licensed under the GPLv3. It is available online at the location given in Ref. [8] and briefly described in Ref. [6].

CAT performs all described methods and some others in a fully automatic procedure, requiring only the multivariate data set as input. Further, a correlation report file is automatically generated that provides all the results. Given the results from the hypothesis test for independence, an experimentalist can make a reliable statement on the dependence of observables in a given data set.

In practice it is recommended to start with e. g. a simulated signal or background data sample with about the same statistics as expected in real data. If dependencies with more than 5σ significance are detected, they definitely should be modeled with conditional PDFs or treated somehow differently. Dependencies with more than 3σ should be studied in more detail, e. g. by increasing the simulated data statistics. If the dependence becomes more significant it is necessary to validate in toy studies if neglecting the dependencies introduces a bias on the analysis. Dependencies below 3σ significance are usually not visible in projections in subranges and are likely due to statistical fluctuations. Thus assuming independence is a valid assumptions.

Increasing the simulated data statistic to e. g. 100 times the statistic of real data should be avoided. Dependencies occurring with this statistic are negligible for the analysis on real data statistic. Further, it might be questionable if the simulation has the proper level of accuracy to describe the dependence to that detail.

Part of the CAT package is also an example that creates a multivariate data set with six partially dependent observables. The observables are generated from uniform U or Gaussian G distributions. The observables are defined as:

$$\begin{aligned} &1 : a = U(0, 1) && 4 : d = U(a, 1) \\ &2 : b = G(0, 1) && 5 : e = r \times \cos\phi \\ &3 : c = a \times b && 6 : f = r \times \cos\phi \end{aligned}$$

with $r = G(0.7, 0.15)$ and $\phi = U(0, 2\pi)$. In Fig. 4.4a the matrix of linear correlations coefficients among the six observables is shown, whereas in Fig. 4.4b the significance for dependence is shown in units of standard deviations. The hypothesis test reveals all dependencies with more than 8σ whereas the linear correlation coefficient deviates from zero only for the pairs of observables $(1, 4)$ and $(2, 3)$. The yellow colored 2.32σ dependence between the pair $(1, 2)$ is a typical example of a statistical fluctuation.

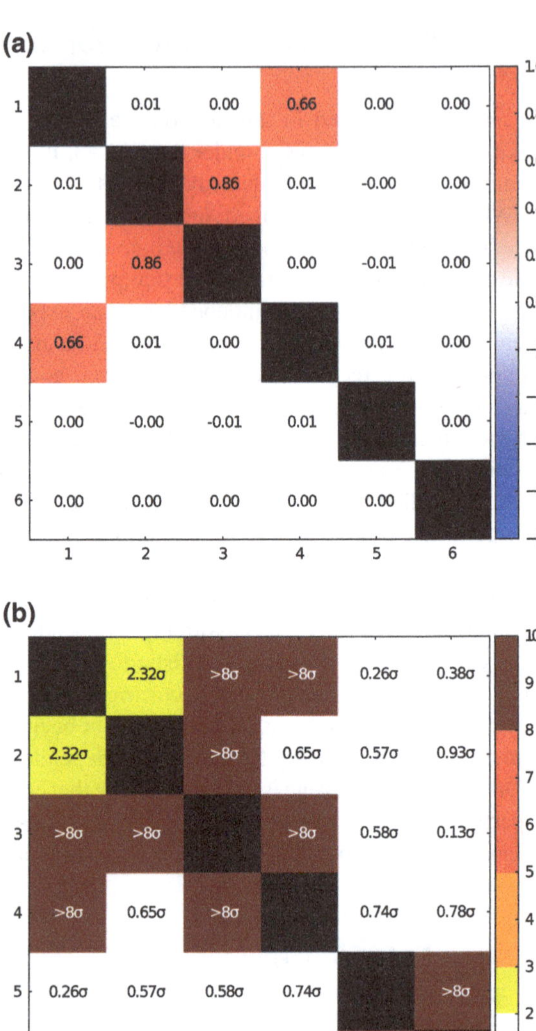

Fig. 4.4 Matrices of **a** linear correlation coefficients and **b** significance for dependence in units of standard deviations for the six observables described in the text

4.4 Error Propagation

The 26 real parameters in Table 2.1 are calculated from the amplitude weights obtained in the ML fit. The uncertainties of some, such as the phases, are rather easy to calculate by using Gaussian error propagation and the correlation matrix of the fitted parameters, as described in Ref. [4, Sect. 4.9]. Although the derivatives, required for the Gaussian error propagation, can always be calculated even if the task may become complicated for certain quantities. A simple and robust method to obtain the correctly propagated statistical uncertainties applies toy based error propagation and avoids calculation of analytic derivatives.

1. Generate a vector $\vec{R}$ of n random Gaussian numbers with mean zero and width one, where n is the number of fitted parameters.
2. Rotate the vector of random numbers, such that they have the same correlation as the fitted parameters by:

   (a) Decompose the correlation matrix C of fitted parameters, using Cholesky decomposition [4, Sect. 3.6], to obtain a matrix U, which is defined as $U^T U = C$.
   (b) The vector of rotated random numbers $\vec{R}_C$ is then given by $\vec{R}_C = U^T \vec{R}$.

3. Scale each rotated random number i in $\vec{R}_C$ by multiplying it with the corresponding statistical uncertainty σ_i, where σ_i is the square root of the ith diagonal element in the covariance matrix of the ML fit.
4. Add the obtained vector of rotated and scaled random numbers to the nominal ML fit result. The obtained result differs from the nominal result by a random value that is consistent with the expectation from the statistical uncertainty.
5. Use the obtained result to calculate all parameters of interest, which, again, differ randomly from the values calculated using the nominal fit result.
6. Repeat the steps 1 to 5 to obtain a distribution for each parameter of interest. The distributions follows a Gaussian distribution and the width corresponds to the statistical uncertainty of the given parameter. Typically, 1,000 to 10,000 repetitions are sufficient to obtain a smooth distribution that can be fitted with a Gaussian distribution with negligible uncertainty on the width.

In case of asymmetric statistical uncertainties, the scaling in step 3 can be done separately with the lower (higher) statistical error for random numbers below (above) zero. The distribution in step 6 can then be fitted, using an asymmetric Gaussian function, to determine the lower (higher) width independently and to obtain the propagated lower (higher) parameter uncertainty.

4.5 Continuum Suppression

4.5.1 Event Shapes

The major background in this analysis originates from $e^+e^- \rightarrow q\bar{q}$ ($q \in \{u, d, s, c\}$) processes. The quark pairs produced in the e^+e^--annihilation fragment into light hadrons and give rise to the dominant source of background, referred to as continuum background.

In these continuum events, the light quark pairs are created back-to-back in the center-of-mass frame and their kinetic energy corresponds almost to the accelerator energy. The hadrons produced in the fragmentation have only small momentum

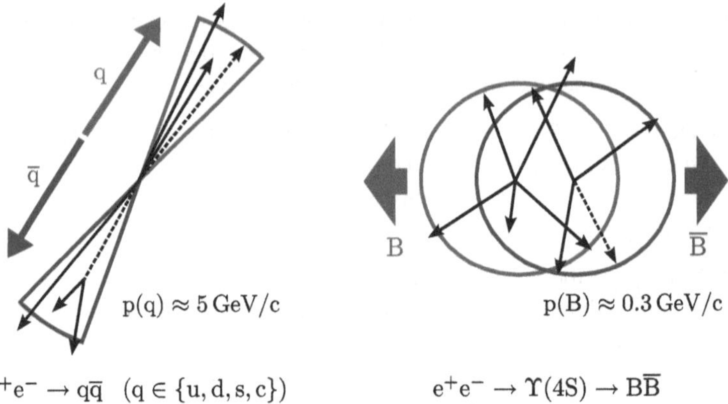

Fig. 4.5 Illustration of the event shapes for continuum (*left*) and $B\bar{B}$ (*right*) events. Light quark pairs in continuum events are produced back-to-back and give rise to a more jet-like structure, whereas $B\bar{B}$ events have a *spherical* shape. Taken from Ref. [9]

perpendicular to the quark flight direction, thus resulting in a jet-like and spatially confined structure. This is in contrast to $B\bar{B}$ events from $e^+e^- \rightarrow \Upsilon(4S) \rightarrow B\bar{B}$ processes. The mass of the $B\bar{B}$ meson pair corresponds almost to the beam energy, therefore it is created nearly at rest in the center-of-mass frame. The pseudoscalar B mesons have spin 0 and thus their decay products have no preferred direction, resulting in an isotropic distribution of spherical shape. The event shapes for continuum and $B\bar{B}$ events are illustrated in Fig. 4.5.

Different quantities can be used to characterize the event shape and describe the event topology. Below the quantities used in this analysis are described in detail. They are further combined in an artificial neural network to construct a more powerful discriminant, which is used in the analysis and described in Sect. 4.5.2.

B Meson Flight Direction with Respect to the Beam

The $\Upsilon(4S)$ is a vector meson with spin 1 and its decay into two pseudo-scalar B mesons with spin 0 requires conservation of angular momentum. In the center-of-mass frame, the polar angle $\cos\theta_B$ of correctly reconstructed B meson candidates with respect to the beam axis follows a $1-\cos^2\theta_B$ distribution. For random combinations of tracks in continuum events the distribution is uniform. In Fig. 4.6a the $|\cos\theta_B|$ distribution of $B\bar{B}$ and continuum events is shown.

Thrust

The thrust T is defined as

$$T = \max_{\vec{T}} = \frac{\sum_i |\vec{p}_i \cdot \vec{T}|}{\sum_i |\vec{p}_i|}, \tag{4.13}$$

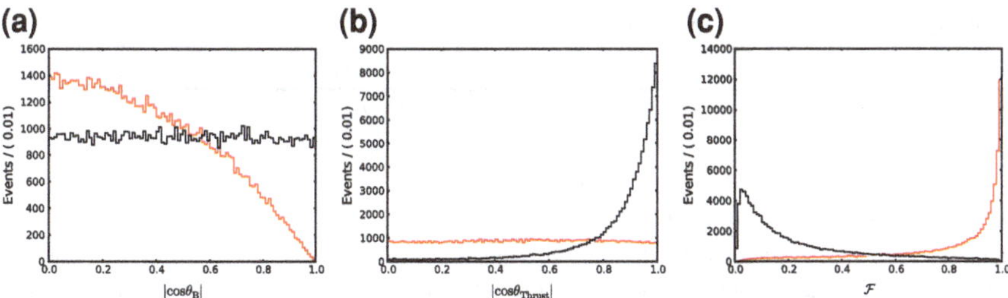

Fig. 4.6 Distribution of **a** $|\cos\theta_{\mathrm{B}}|$, **b** $|\cos\theta_{\mathrm{Thrust}}|$, and **c** $\mathcal{F}$ for MC simulated $\mathrm{B\bar{B}}$ (*red*) and continuum (*black*) events

where $\vec{p}_i$ is the momentum of the ith final-state particle in an event and $\vec{T}$ is the thrust axis, defined as the direction that maximises the sum of the longitudinal momenta of particles. Final-state particles are the stable decay products that are measured in the different detector components, described in Sect. 3.2. The concept of thrust was originally introduced to quantify jets at high energy experiments, back in the 1980s [10].

An observable that can be used to distinguish between continuum and $\mathrm{B\bar{B}}$ events is $|\cos\theta_{\mathrm{Thrust}}|$, where θ_{Thrust} is the angle between the thrust axis of the reconstructed B candidate and the thrust axis of all remaining particles in the event. For $\mathrm{B\bar{B}}$ events the $|\cos\theta_{\mathrm{Thrust}}|$ distribution is expected to be uniform as the isotropically distributed particles in the event have random thrust axes. For the jet-like continuum events it is more likely that the thrust axes are collinear. If the reconstructed B candidate is combined by particles of one jet the $|\cos\theta_{\mathrm{Thrust}}|$ distribution is peaking towards 1. The $|\cos\theta_{\mathrm{Thrust}}|$ distribution of $\mathrm{B\bar{B}}$ and continuum events is shown in Fig. 4.6b.

Fox-Wolfram Moments and Modified Fox-Wolfram Moments

The kth Fox-Wolfram moment H_k and the kth normalized Fox-Wolfram moment R_k are defined as

$$H_k = \sum_{ij} \frac{|\vec{p}_i||\vec{p}_j| P_k\left(\cos\theta_{ij}\right)}{E_{\mathrm{vis}}^2} \quad \text{and} \quad R_k = \frac{H_k}{H_0}, \tag{4.14}$$

where P_k denotes the kth Legendre polynomial, $\vec{p}_i$ ($\vec{p}_j$) is the momentum of the ith (jth) final-state particle, θ_{ij} is the opening angle between the momenta of the ith and jth particle and E_{vis} is the sum of the measured energy in the event. The Fox-Wolfram moments have been introduced to describe event shapes in $\mathrm{e^+e^-}$-annihilations [11, 12].

In the study of charmless Bdecays, the Belle Collaboration introduced modified Fox-Wolfram moments [13, 14], sometimes also referred to as Super-Fox-Wolfram moments, which are a refinement of the Fox-Wolfram moments. The summation over

Table 4.1 Input variables of the neural network

Rank	Variable	Added (σ)	Individual (σ)	Global correl. (%)
1	$\mathcal{F}$	292.67	292.67	73.8
2	$\lvert\cos\theta_{\mathrm{B}}\rvert$	51.59	94.36	15.0
3	$\lvert\cos\theta_{\mathrm{Thrust}}\rvert$	46.90	247.77	73.5
Total signi.: 300.86σ Total correl. to target: 69.81 %				

Rank, variable name, added and individual significance of the variable, and global correlation are listed for each variable. Total significance of the network and correlation of the network output to the target (signal) are given

the ith and jth particle momenta in Eq. (4.14) is not taken over all final-state particles in the event. Instead, the summation is taken over the final-state particles of the reconstructed B candidate, denoted by the superscript "s", or over the remaining final-state particles in the event, denoted by the superscript "o". The possible combinations give rise to three groups of modified Fox-Wolfram moments R_k^{so}, R_k^{oo} and R_k^{ss}.

The moments R_k^{so} and R_k^{oo} can be used to distinguish between continuum and $\mathrm{B\bar{B}}$ events. The moments R_k^{ss} as well as R_1^{so} and R_3^{so} are generally excluded due to correlations with M_{bc} and ΔE, two observables used in this analysis and a multitude of other analyses at the Belle experiment. Additional improvement is achieved by performing the calculation of R_k^{so} in categories that represent the quality of reconstructed events based on the missing mass in the event.

The different moments have linear dependence among each other and are combined using a Fisher discriminant [15] to obtain a single discriminating observable $\mathcal{F}$ with values from 0 (for jet-like continuum events) to 1 (for spherical $\mathrm{B\bar{B}}$ events). The distribution of $\mathcal{F}$ is shown in Fig. 4.6c.

4.5.2 Artificial Neural Network

The three discriminating observables $\lvert\cos\theta_{\mathrm{B}}\rvert$, $\lvert\cos\theta_{\mathrm{Thrust}}\rvert$, and $\mathcal{F}$ are further combined to a super-discriminant C_{NB} using an artificial neural network that can handle also potential non-linear dependence among the three observables. The neural network is realised with the NeuroBayes package [16, 17], which provides, beside the training of neural networks itself, advanced algorithms for preprocessing and regularization of input data. The package makes internal use of the Bayes theorem [18] and has proven its performance and robustness at various high energy physics experiments and in industrial applications.

The training of the neural network was performed with MC simulated continuum events for background and MC simulated $\mathrm{B}^0 \to \phi(\mathrm{K}^+\pi^-)^*$ events for signal (see also Sect. 3.3). The signal events were simulated unpolarized as a sum of $\mathrm{B}^0 \to \phi(\mathrm{K}\pi)_0^*$, $\mathrm{B}^0 \to \phi\mathrm{K}^*(892)^0$, and $\mathrm{B}^0 \to \phi\mathrm{K}_2^*(1430)^0$ decays with relative strengths given by the latest branching fraction measurements in Ref. [5].

The observables used as input variables to the network are given in Table 4.1. The input variables are sorted by their rank, their power to separate between signal

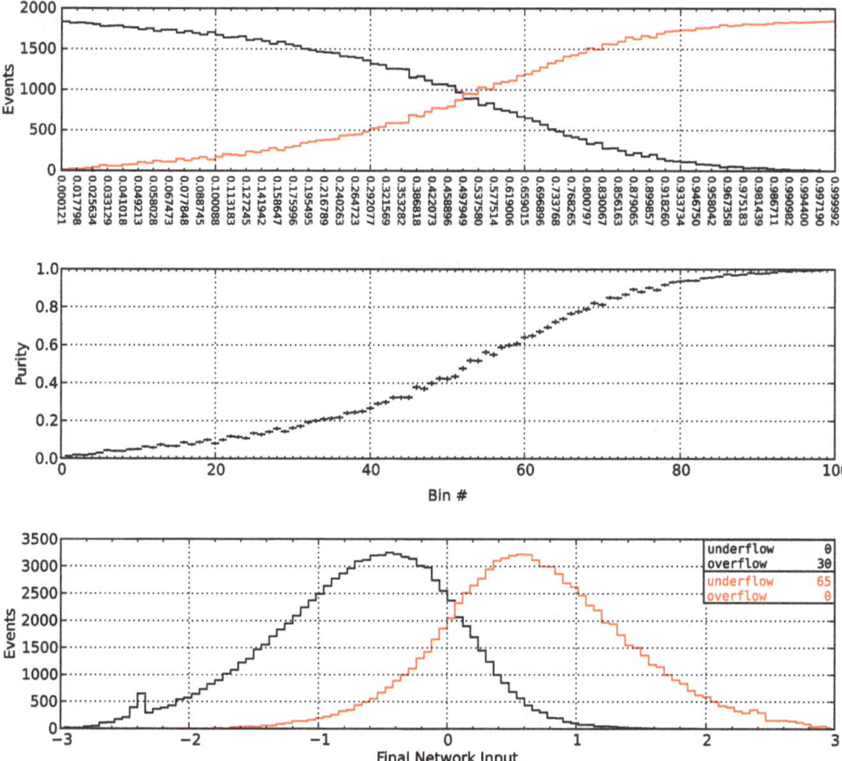

Fig. 4.7 Flattened distribution of $\mathcal{F}$ (*top*) shown as individual contribution from signal (*red*) and background (*black*), purity of the flattened distribution per bin (*middle*), and purity transformed to Gaussian with mean zero and width one (*bottom*), again shown as individual contribution from signal (*red*) and background (*black*). Values exceeding ± 3 in the *bottom* figure are given as over- and underflow entries

and background after the preprocessing was performed. The added significance is a measure of how much this input variable adds to the separation of the network between signal and background, whereas the individual significance is a measure of how good the variable itself separates. Zero individual significance implies identical distributions for signal and background. The global correlation is a measure of how strong the variable is correlated to others. Large values typically result in low values for the added significance, except for one variable that is ranked high. The total significance and total correlation to target are measures for the networks ability to separate between signal and background. Again, zero significance or zero correlation corresponds to the inability of the network to separate among them. A definition of significance can be found in the section about significance tests in Ref. [5, Sect. 36.2.2].

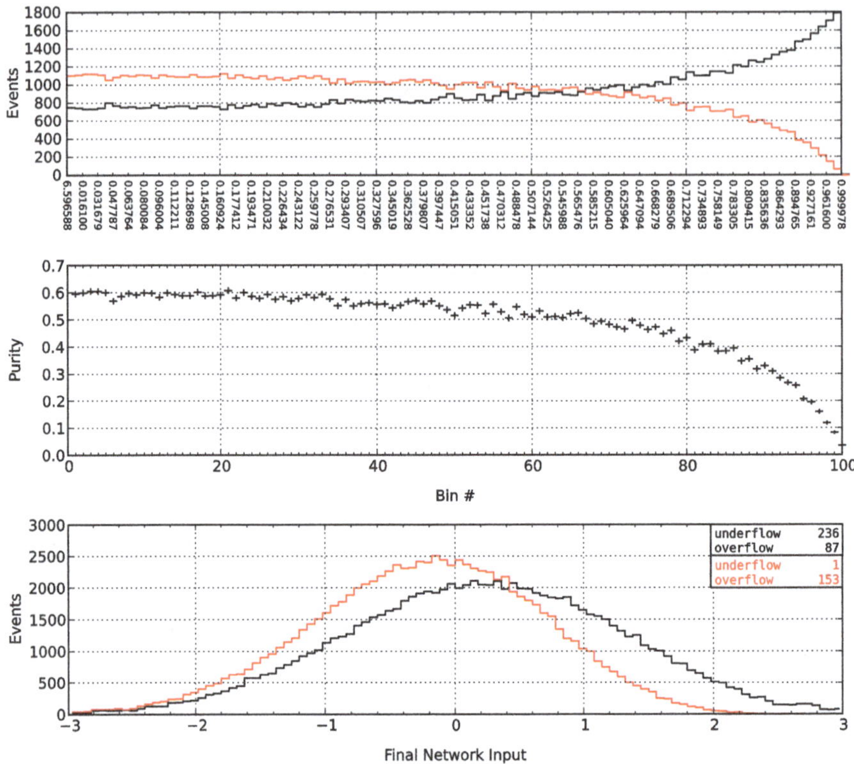

Fig. 4.8 Flattened distribution of $|\cos\theta_{\mathrm{B}}|$ (*top*) shown as individual contribution from signal (*red*) and background (*black*), purity of the flattened distribution per bin (*middle*), and purity transformed to Gaussian with mean zero and width one (*bottom*), again shown as individual contribution from signal (*red*) and background (*black*). Values exceeding ± 3 in the *bottom* figure are given as over- and underflow entries

The preprocessing applied in this analysis flattens the input distribution and transforms it to a Gaussian with mean zero and width one. Other types of preprocessing exist for example for discrete input variables or input variables with missing values. A global preprocessing is applied which accepts only variables with at least 3σ added significance in the training and decorrelates and rotates the input variables to improve the network training. Control plots from the preprocessing during the neural network training are given for $\mathcal{F}$, $|\cos\theta_{\mathrm{B}}|$, and $|\cos\theta_{\mathrm{Thrust}}|$ in Figs. 4.7, 4.8 and 4.9, respectively.

The resulting C_{NB} distribution of signal and background events on the training sample is shown in Fig. 4.10a. In many analyses, the network output is used to reject background events by requiring it to be larger than a certain "cut" value, i.e. $C_{\mathrm{NB}} > \mathrm{cut}$. More sophisticated use of the information in the network output can be made by including it as an additional observable in a multidimensional ML fit.

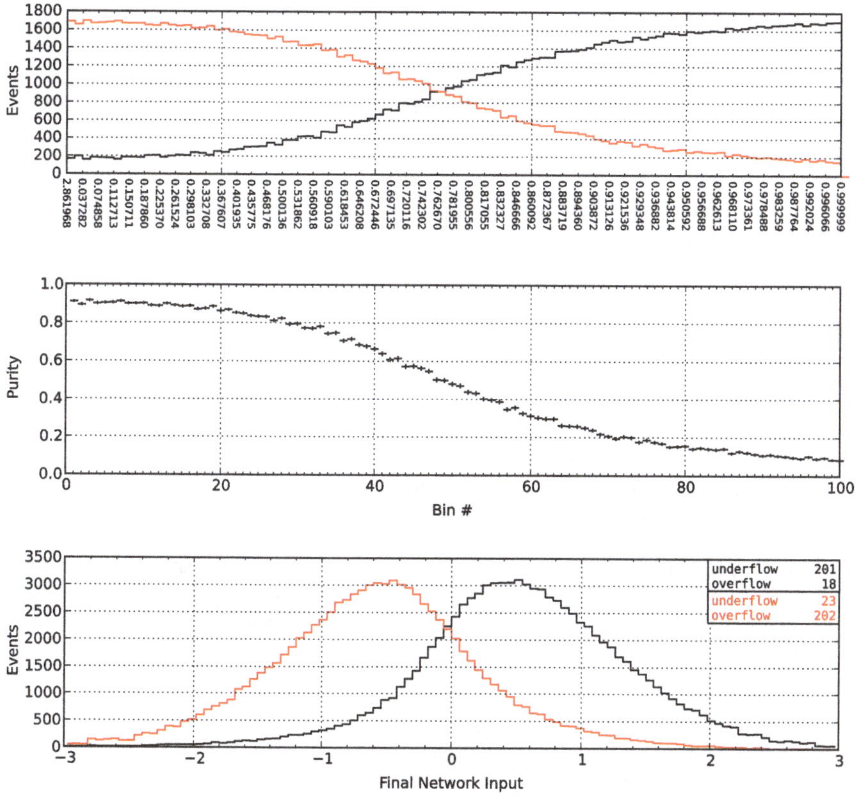

Fig. 4.9 Flattened distribution of $|\cos\theta_{\mathrm{Thrust}}|$ (*top*) shown as individual contribution from signal (*red*) and background (*black*), purity of the flattened distribution per bin (*middle*), and purity transformed to Gaussian with mean zero and width one (*bottom*), again shown as individual contribution from signal (*red*) and background (*black*). Values exceeding ±3 in the *bottom* figure are given as over- and underflow entries

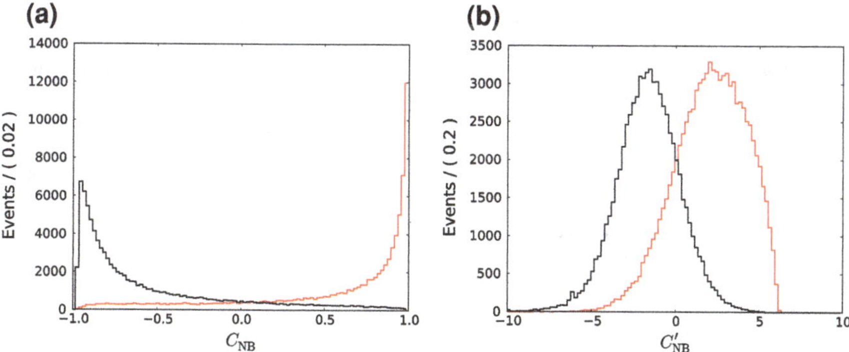

Fig. 4.10 Distributions of **a** the network output C_{NB} and **b** the transformed network output C'_{NB} with a requirement of $C_{\mathrm{NB}} > -0.9$ applied in (**b**) for signal (*red*) and background (*black*)

Typically, a soft cut is performed on C_{NB} and the remaining events are transformed to a new observable

$$C'_{\mathrm{NB}} = \log\left(\frac{C_{\mathrm{NB}} - \mathrm{cut}}{1 - C_{\mathrm{NB}}}\right), \tag{4.15}$$

where C'_{NB} tends to have a Gaussian-like shape that can usually be described analytically by a sum of Gaussian functions. An exemplary distribution of C'_{NB} is shown in Fig. 4.10b with a requirement of $C_{\mathrm{NB}} > -0.9$ applied.

References

1. F. James et al., *MINUIT, Function Minimization and Error Analysis* (CERN Program Library, Geneva)
2. W. Verkerke, D. Kirkby, The RooFit toolkit for data modeling. in *Proceedings from the Computing in High Energy and Nuclear Physics 2003 Conference*, `arXiv:physics/0306116`
3. R. Brun, F. Rademakers, ROOT: an object oriented data analysis framework. Nucl. Instr. Meth. A **389**, 81–86 (1997)
4. V. Blobel, E. Lohrmann, Statistische und numerische Methoden der Datenanalyse. Teubner Verlag (1998), http://www.desy.de/~blobel/eBuch.pdf
5. J. Beringer et al., (Particle Data Group), The review of particle physics. Phys. Rev. D **86**, 010001 (2012), http://pdg.lbl.gov
6. M. Feindt, M. Prim, An algorithm for quantifying dependence in multivariate data sets. Nucl. Instr. Meth. A **698**, 84–89 (2013)
7. R.B. Nelson, *An Introduction to Copulas* (Springer, New York, 2006)
8. CAT—A correlation analysis toolkit, https://ekptrac.physik.uni-karlsruhe.de/trac/CAT
9. M. Röhrken, Time-dependent CP violation measurements in neutral B Meson to double-charm decays at the Japanese Belle experiment. PhD thesis, Institute for Experimental Nuclear Physics, Karlsruhe Institute of Technology, IEKP-KA/2012-13 (2012)
10. E. Farhi, Quantum chromodynamics test for jets. Phys. Rev. Lett. **39**, 1587–1588 (1977)
11. G.C. Fox, S. Wolfram, Observables for the analysis of event shapes in e^-e^+ annihilation and other processes. Phys. Rev. Lett. **41**, 1581–1585 (1978)
12. G.C. Fox, S. Wolfram, Event shapes in e^-e^+ annihilation. Nucl. Phys. B **149**(3), 413–496 (1979)
13. K. Abe et al., (Belle Collaboration), Measurement of branching fractions for $B \to \pi\pi$, $K\pi$ and KK decays. Phys. Rev. Lett. **87**, 101801 (2001)
14. K. Abe et al., (Belle Collaboration), A measurement of the branching fraction for the Inclusive $B \to X_s\gamma$ decays with the Belle detector. Phys. Lett. B **511**, 151–158 (2001)
15. R.A. Fisher, The use of multiple measurements in taxonomic problems. Ann. Eugenics **7**(2), 179–188 (1936)
16. M. Feindt, A neural Bayesian estimator for conditional probability densities (2004), `arXiv:physics/0402093`
17. M. Feindt, U. Kerzel, The NeuroBayes neural network package. Nucl. Instr. Meth. A **559**, 190 (2006)
18. T. Bayes, An essay towards solving a problem in the doctrine of chances. Philos. Trans. R. Soc. B **53**, 370–418 (1763)

Chapter 5
Reconstruction and Selection

This chapter addresses the reconstruction of B^0 mesons and the applied selection criteria. It further discusses different background contributions and the reconstruction efficiency of signal events. Finally, a control channel for different cross-checks throughout the analysis is introduced.

5.1 Event Reconstruction and Selection

The B^0 meson candidates are reconstructed in the exclusive decay chain $B^0 \to \phi K^* \to (K^+K^-)(K^+\pi^-)$. The reconstruction takes the reconstructed charged tracks and combines them to ϕ and K^* candidates, which are further combined to B^0 candidates. The applied selection criteria are chosen to include as many candidates as possible in the final ML fit. The reconstruction and selection was developed and optimized with MC simulated data events. Below, a summary of each step of the reconstruction chain is given.

5.1.1 Charged Tracks

The charged tracks are required to have a transverse (longitudinal) distance of closest approach to the interaction point of less than 0.1 (4.0) cm. Information from the CDC, ACC, and TOF subdetectors, described in Sect. 3.2, are combined in a likelihood-ratio and provide a particle identification (PID) quantity for each track.

5.1.2 ϕ Mesons

The ϕ meson candidates are reconstructed using two oppositely charged tracks with a K^+K^- invariant mass requirement of $M_{KK} < 1.05$ GeV. The requirements on the

M. Prim, *Polarization and CP Violation Measurements*, Springer Theses,
DOI: 10.1007/978-3-319-05756-9_5,

PID quantity for these two tracks provide a kaon identification efficiency of 95 % with an associated pion misidentification rate of 26 %.

5.1.3 B^0 *Mesons*

The K^* meson candidates are reconstructed using two oppositely charged tracks with a $\mathrm{K}^+\pi^-$ invariant mass criterion of $0.7\,\mathrm{GeV} < M_{K\pi} < 1.55\,\mathrm{GeV}$ applied. The PID requirement for the K^+ track candidate has a kaon identification efficiency of 90 % with an associated pion misidentification rate of 28 %, whereas the PID requirement for the π^- candidate has a pion identification efficiency of 98 % with an associated kaon misidentification rate of 9 %.

5.1.4 *Self-Crossfeed Background*

The ϕ and K^* candidates are combined to B^0 meson candidates. Two quantities are used for the B^0 selection: the beam-energy-constrained mass

$$M_{\mathrm{bc}} = \sqrt{(E^*_{\mathrm{beam}})^2 - (p^*_B)^2} \tag{5.1}$$

and the energy difference

$$\Delta E = E^*_B - E^*_{\mathrm{beam}}, \tag{5.2}$$

where E^*_{beam} is half of the beam energy, and p^*_B and E^*_B are the momentum and energy of the B^0 candidate in the center-of-mass frame, respectively. Candidates in the region $5.24\,\mathrm{GeV} < M_{\mathrm{bc}} < 5.29\,\mathrm{GeV}$ and $-150\,\mathrm{MeV} < \Delta E < 150\,\mathrm{MeV}$ are retained for further analysis. The M_{bc} range is divided into two regions: a sideband for candidates in the subrange $5.24\,\mathrm{GeV} < M_{\mathrm{bc}} < 5.26\,\mathrm{GeV}$ and the nominal fit region in the subrange $5.26\,\mathrm{GeV} < M_{\mathrm{bc}} < 5.29\,\mathrm{GeV}$.

In 17 % of all signal events, more than one B^0 candidate passes all selection criteria. In that case, all tracks of one candidate are constrained to originate from a common vertex within the interaction region. The candidate with the smallest χ^2 for this hypothesis is kept. This requirement selects the correct candidate with a probability of 64 %, according to MC simulations.

Two more selection requirements, on the neural network output C_{NB} and on the helicity angle $\cos\theta_1$, are further applied to reject background events as described in the next section.

5.2 Background Studies

5.2.1 Combinatorial Background

The selected sample of B^0 candidates is dominated by background candidates from $e^+e^- \rightarrow q\bar{q}$ ($q \in \{u, d, s, c\}$) events, in which random combinations of tracks pass all selection criteria and fake a signal candidate. The neural network output C_{NB}, described in Sect. 4.5.2, must satisfy the criterion $C_{NB} > 0$, which rejects 86 % of this continuum background and retains 83 % of the signal.

Another background component, with a small fraction of about 2 % of the continuum background, arises from random combination of tracks in $B\overline{B}$ events. These events follow a similar distribution as the continuum background events in all observables and are indistinguishable in data. In Fig. 5.1a the M_{bc} distribution of MC simulated data events is shown and illustrates the dominating background from continuum events and the small contribution from $B\overline{B}$ events. Hereinafter, the sum of both components is referred to as combinatorial background.

5.2.2 Self-Crossfeed Background

In some signal events, the B^0 candidate is reconstructed only partially with one or more tracks originating from the other B meson in the event. The wide $K^+\pi^-$ invariant mass window of $0.7\,\text{GeV} < M_{K\pi} < 1.55\,\text{GeV}$ allows for random combinations that still pass the criterion. As about 70 % of all tracks in a $B\overline{B}$ event are pions, it is most likely that a random π^- track from the other Bmeson is used to reconstruct the signal candidate. Typically, the momentum in the lab frame of such random π^- is smaller than the K^+ momentum which results in the $K^+\pi^-$ system being dominated by the K^+ momentum. Therefore, these self-crossfeed (SCF) events tend to peak in the region of high $\cos\theta_l$ values, as illustrated in Fig. 5.1b. A part of the SCF events can be rejected, see next section, and the remaining events are treated as a systematic uncertainty.

5.2.3 Peaking Background

Certain B meson decay chains can fake signal events. The inclusive $b \rightarrow c$ and $b \rightarrow s$ MC samples with four and 50 times the statistics expected in data, respectively, are used to study such decays. The branching fractions in the MC simulation have been scaled to match the latest measurements from Ref. [1]. Below, an overview is given containing either decay chains that have the same final state or one misidentified track. Decay chains with more than one track being misidentified have been found to be negligible.

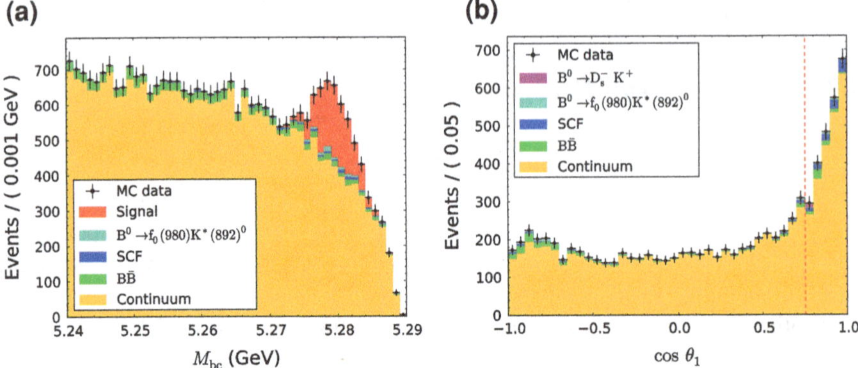

Fig. 5.1 Distributions of MC simulated data events in **a** $M_{\rm bc}$ and **b** $\cos\theta_1$. In **a** and **b** the requirement $C_{\rm NB} > 0$ is applied, yet the continuum background is still dominating. In **a** the requirement on $\cos\theta_1 < 0.75$ is applied. In **b** correctly reconstructed signal events are not shown as their distribution depends on the polarization and $M_{\rm bc} > 5.27$ is required to enrich the peaking background contributions. The vertical line in **b** indicates the $\cos\theta_1 < 0.75$ requirement

Two decay modes,

1. $\mathrm{B^0 \to D_s^- K^+ \to (\phi\pi^-)K^+ \to ([K^+K^-]\pi^-)K^+}$
2. $\mathrm{B^0 \to f_0(980)K^*(892)^0 \to (K^+K^-)(K^+\pi^-)}$

show a clearly peaking structure in at least the $M_{\rm bc}$, ΔE, and M_{KK} distribution with a statistically significant amount of events expected from MC simulations. However, the $\mathrm{B^0 \to D_s^- K^+}$ mode is a series of two-body decays and peaks sharply near 0.8 in the $\cos\theta_1$ distribution, as can be seen in Fig. 5.1b. A requirement of $\cos\theta_1 < 0.75$ is applied to reject the peaking component entirely as well as the majority of SCF events; both can be seen in Fig. 5.1a. The requirement further rejects a large part of the combinatorial background.

The $\mathrm{B^0 \to f_0(980)K^*(892)^0}$ mode denotes any contribution from a broad scalar component in the $\mathrm{K^+K^-}$ distribution that could originate from either $\mathrm{f_0(980)}$, $\mathrm{a_0(980)}$, or non-resonant $\mathrm{K^+K^-}$. This mode can not be vetoed and is included as an additional component in the ML fit. There is no statistically significant contribution expected from $\mathrm{K^*}$ states other than the $\mathrm{K^*(892)^0}$ but the possibility is checked as a systematic uncertainty.

Five decay modes,

1. $\mathrm{B^0 \to \phi\phi \to (K^+K^-)(K^+K^-)}$
2. $\mathrm{B^0 \to \phi\rho^0 \to (K^+K^-)(\pi^+\pi^-)}$
3. $\mathrm{B^0 \to f_0(980)\pi^+\pi^- \to (K^+K^-)\pi^+\pi^-}$
4. $\mathrm{B^0 \to \phi f_0(980) \to (K^+K^-)(\pi^+\pi^-)}$
5. $\mathrm{B^0 \to \phi f_0(980) \to (K^+K^-)(K^+K^-)}$

that could be identified as possible peaking backgrounds, are all unobserved and only upper limits on their branching fractions are given in Ref. [1]. Assuming these upper limits to be branching fractions, no statistically significant contribution that would effect the ML fit is expected from MC simulations. Due to one misidentified track, all modes show displaced peaks in at least the ΔE distribution and peak either several MeV above or below the signal. This depends on the mass difference of the different particle hypotheses. These modes are not included in the ML fit but are studied as a possible systematic uncertainty.

5.3 Reconstruction Efficiency

The general detector acceptance and the selection requirements described in the previous sections affect the reconstruction efficiency. Whereas for observables such as $M_{\rm bc}$ the true distribution has a negligible width and the observed distribution is only smeared by the detector resolution, the distributions of the mass-angular observables $M_{K\pi}$, $\cos\theta_1$, $\cos\theta_2$, and Φ require a more detailed study.

The signal distributions in the four mass-angular observables are broad and depend on the polarization of the $\mathrm{B}^0 \to \phi\mathrm{K}^*$ system. For a $\mathrm{B}^0 \to \phi\mathrm{K}^+\pi^-$ three-body phase space decay with uniform angular distributions, the averaged reconstruction efficiency is found to be $(28.3 \pm 0.1)\%$, according to MC simulation. In general, for a given partial wave with spin J and non-uniform acceptance, the reconstruction efficiency $\epsilon_{\mathrm{reco},J}$ depends on the observed mass-angular distribution and can be obtained only after the polarization was measured. For the partial wave amplitudes $\mathcal{A}_J$ in Eqs. (2.23–2.25), one can compute $\epsilon_{\mathrm{reco},J}$ using

$$\epsilon_{\mathrm{reco},J} = \frac{\int \alpha\left(M_{K\pi}, \cos\theta_1, \cos\theta_2, \Phi\right) |\mathcal{A}_J|^2}{\int |\mathcal{A}_J|^2} = \frac{n}{d}, \tag{5.3}$$

where $\alpha\left(M_{K\pi}, \cos\theta_1, \cos\theta_2, \Phi\right)$ is a four-dimensional acceptance function and the numerator n is the integral over the phase space with the acceptance included and is given by

$$n = \int_{-\pi}^{\pi}\int_{-1}^{+1}\int_{-1}^{+1}\int_{m_K+m_\pi}^{m_{B^0}-m_\phi} \alpha|\mathcal{A}_J|^2 dM_{K\pi} d\cos\theta_1 d\cos\theta_2 d\Phi, \tag{5.4}$$

where m_{B^0}, m_ϕ, m_K, and m_π are the nominal particle masses from Ref. [1] that limit the $M_{K\pi}$ phase space. The explicit dependencies of α and $\mathcal{A}_J$ are omitted for readability. The denominator of Eq. (5.3), d, is given by the integral over the full phase space with a uniform acceptance

$$d = \int_{-\pi}^{\pi}\int_{-1}^{+1}\int_{-1}^{+1}\int_{m_K+m_\pi}^{m_{B^0}-m_\phi} |\mathcal{A}_J|^2 dM_{K\pi} d\cos\theta_1 d\cos\theta_2 d\Phi. \tag{5.5}$$

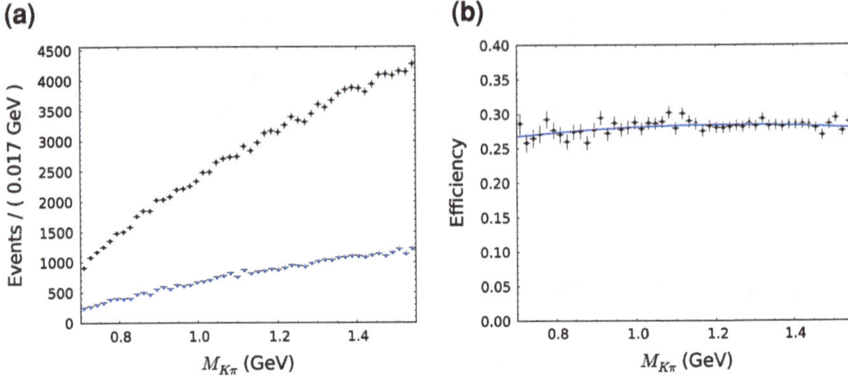

Fig. 5.2 Efficiency as a function of $M_{K\pi}$ with **a** generated (*black marker*) and observed (*blue marker*) distribution of events as a function of $M_{K\pi}$ and **b** resulting efficiency (*black marker*) and parametrized acceptance function (*blue line*)

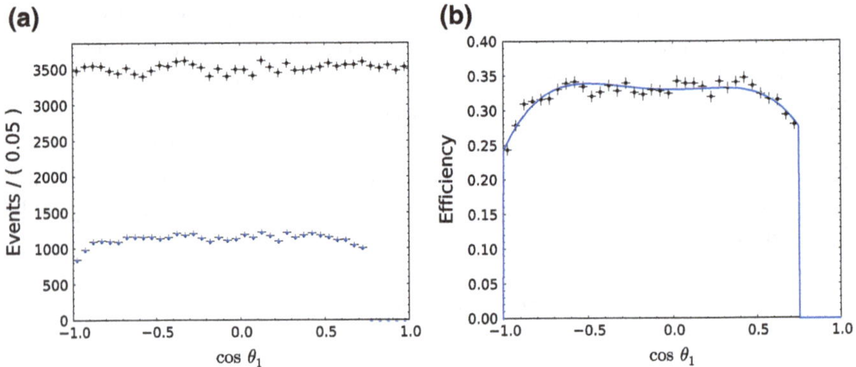

Fig. 5.3 Efficiency as a function of $\cos\theta_1$ with **a** generated (*black marker*) and observed (*blue marker*) distribution of events as a function of $\cos\theta_1$ and **b** resulting efficiency (*black marker*) and parametrized acceptance function (*blue line*)

The four-dimensional mass-angular acceptance function $\alpha\big(M_{K\pi}, \cos\theta_1, \cos\theta_2, \Phi\big)$ is modeled by the product of the four properly normalized one-dimensional efficiency functions. In Figs. 5.2–5.5 the efficiency as a function of $M_{K\pi}$, $\cos\theta_1$, $\cos\theta_2$, and Φ is shown for a MC simulated $\mathrm{B}^0 \to \phi\mathrm{K}^+\pi^-$ three-body phase space decay, which is generated with uniform angular distributions. The efficiency distribution in $M_{K\pi}$ is parametrized using a second-order polynomial function. The efficiency as a function of $\cos\theta_1$ is parametrized by a forth-order polynomial function for $\cos\theta_1 < 0.75$ and zero above. The efficiency as a function of $\cos\theta_2$ and Φ is found to be uniform.

Polynomial functions with at least two orders higher than described above have been used to parametrize the efficiency as function of each observable, but the corresponding coefficients have been found to be consistent with zero within their statisti-

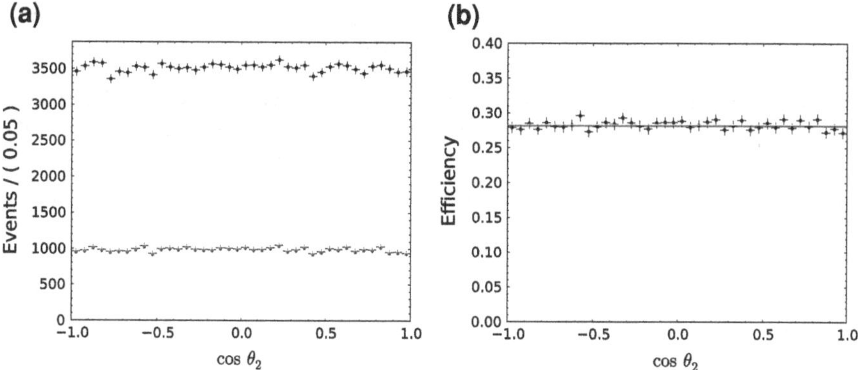

Fig. 5.4 Efficiency as a function of $\cos\theta_2$ with **a** generated (*black marker*) and observed (*blue marker*) distribution of events as a function of $\cos\theta_2$ and **b** resulting efficiency (*black marker*) and parametrized acceptance function (*blue line*)

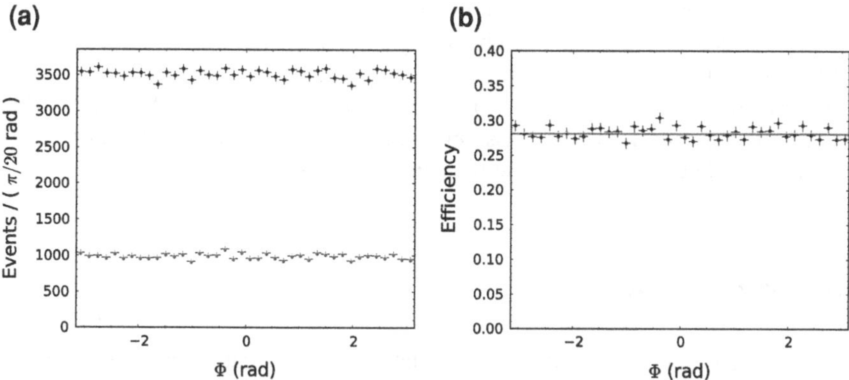

Fig. 5.5 Efficiency as a function of Φ with **a** generated (*black marker*) and observed (*blue marker*) distribution of events as a function of Φ and **b** resulting efficiency (*black marker*) and parametrized acceptance function (*blue line*)

cal uncertainty. A possible flavor dependence in the parametrization, i.e. a difference due to the charge of the primary kaon between $B^0 \to \phi K^+\pi^-$ and $\overline{B}^0 \to \phi K^-\pi^+$ samples, has also been studied, but was found to be smaller than the statistical uncertainty of the parametrization obtained on the joined samples.

To further cross-check the validity of the acceptance function being modeled as product of one-dimensional functions, two-dimensional products are studied. The expected two-dimensional efficiency $\epsilon_{1D\times 1D}$ from the product of one-dimensional functions is compared to the efficiency obtained from a two-dimensional histogram $\epsilon_{2D-Hist}$, using the MC simulated $B^0 \to \phi K^+\pi^-$ three-body phase space sample.

The deviation σ_{eff} of the expected two-dimensional efficiency in units of the statistical uncertainty of the two-dimensional histogram $\sigma_{2D-Hist}$ is given by

$$\sigma_{\text{eff}} = \frac{\epsilon_{1\text{D}\times 1\text{D}} - \epsilon_{2\text{D}-\text{Hist}}}{\sigma_{2\text{D}-\text{Hist}}}. \tag{5.6}$$

For the six possible two-dimensional combinations of the four mass-angular observables, the deviations are shown in Fig. 5.6 with an equidistant binning. For $\cos\theta_1$ the study is limited to the region of $\cos\theta_1 < 0.75$. The mean of deviations is consistent with zero for all six combinations. The χ^2 of the deviations can be computed as

$$\chi^2 = \sum_i^{n_i}\sum_j^{n_j}(\sigma_{\text{eff}}^{(i,j)})^2, \tag{5.7}$$

where $\sigma_{\text{eff}}^{(i,j)}$ is the content of the $(i, j)^{\text{th}}$ bin. Using the number of bins $n_i \times n_j$ as degrees of freedom, the significance of the hypothesis is less than 1σ for all combinations that the distribution of the expected efficiency from the product of one-dimensional functions is different from the efficiency of the two-dimensional histogram, except for the combination of $M_{K\pi}$ and $\cos\theta_1$, shown in Fig. 5.6a. The significance is 3.2σ and dominated by a single bin in the bottom right and becomes 1.1σ if this bin is removed from the χ^2.

Overall, the product of one-dimensional efficiency functions provides a reasonable model of the four-dimensional mass-angular acceptance function. Uncertainties on this model will be included as systematic uncertainties.

5.4 Control Channel

Certain differences between the MC simulation and data can be estimated from a control sample. Control channels are typically channels with higher statistics than the signal channel and a high purity, thus providing a sample to compare the MC simulation of the control channel directly with its corresponding data. As a control channel the decay $\text{B}^0 \to \text{J}/\psi\,\text{K}^*(892)^0 \to (\mu^+\mu^-)(\text{K}^+\pi^-)$ was selected, where the ϕ meson was replaced by a J/ψ meson with respect to the signal channel.

The J/ψ candidates are reconstructed using two oppositely charged tracks with an $\mu^+\mu^-$ invariant mass requirement of $3.085\,\text{GeV} < M_{\mu\mu} < 3.110\,\text{GeV}$. The $\text{K}^*(892)^0$ candidates are reconstructed using two oppositely charged tracks with a $\text{K}^+\pi^-$ invariant mass criterion of $0.85\,\text{GeV} < M_{K\pi} < 0.95\,\text{GeV}$ applied. For the kaon and pion track candidate from $\text{K}^*(892)^0$, the same PID requirements as for the K^* mesons in the signal channel are applied. For the muon candidates the PID is required to be consistent with a muon hypothesis. The energy difference of the B^0 candidates, combined from J/ψ and $\text{K}^*(892)^0$ candidates, is required to be $-20\,\text{MeV} < \Delta E < 20\,\text{MeV}$ and the beam-energy-constrained mass must satisfy the criterion $5.24\,\text{GeV} < M_{\text{bc}} < 5.29\,\text{GeV}$.

In Fig. 5.7 the normalized M_{bc} distribution of the selected B^0 candidates is shown. The data distribution is nearly background free as no combinatorial background from

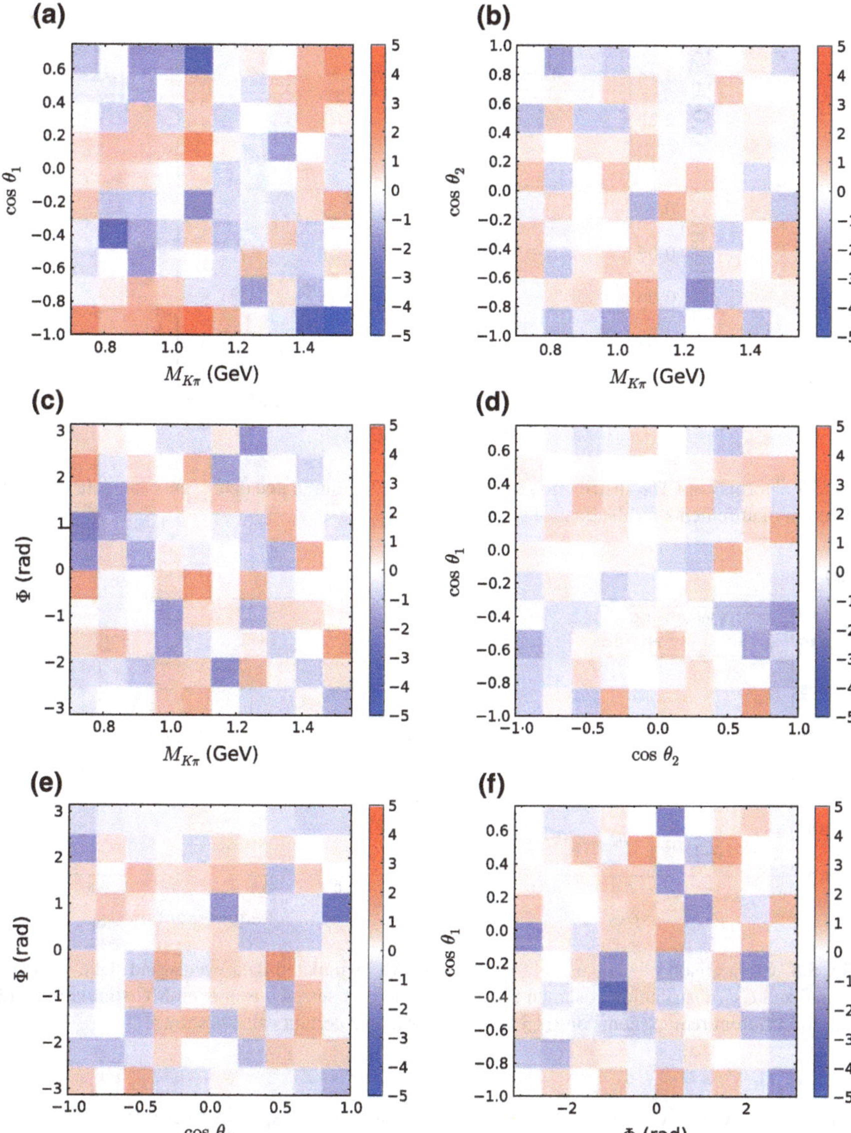

Fig. 5.6 Deviations of the two-dimensional efficiency calculated from the product of one-dimensional functions from the actual two-dimensional efficiency in units of the uncertainty of the efficiency. In **a** to **f** the six possible two-dimensional combinations of the four mass-angular observables $M_{K\pi}$, $\cos\theta_1$, $\cos\theta_2$, and Φ are shown

continuum events is expected due to the tight selection around the narrow J/ψ peak. The only expected background is from self-crossfeed and combinatorial background in $B \to J/\psi X$ inclusive events. The distribution of such J/ψ inclusive MC simulated

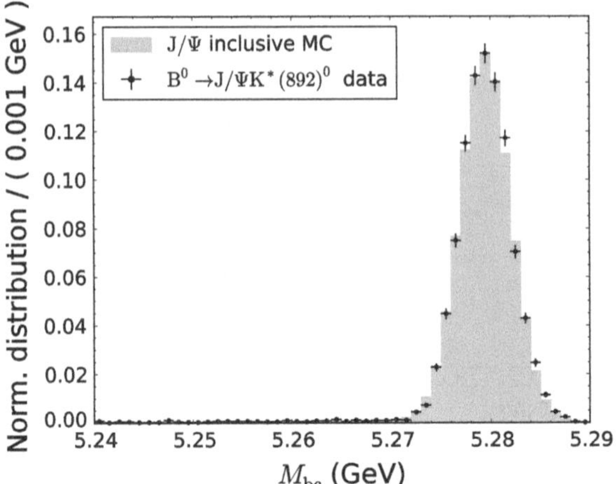

Fig. 5.7 Normalized M_{bc} distribution of J/ψ inclusive MC simulated data events and data with the selection requirements for the control channel applied

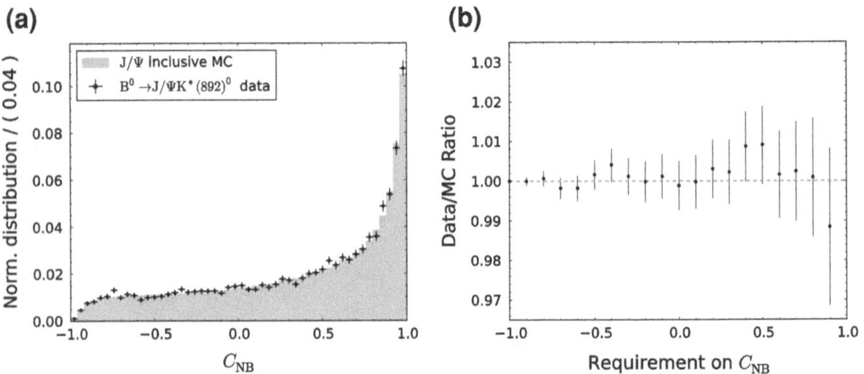

Fig. 5.8 Comparison of C_{NB} using J/ψ inclusive MC simulated data events and data. In **a** the normalized C_{NB} distribution is shown and in **b** the ratio of selected events on MC simulation and data for different requirements on C_{NB} being larger than a certain value is shown

data events, with 50 times the statistics expected on data, is in excellent agreement with the observed data distribution. In the region $M_{bc} > 5.27$ GeV, 9599 events are selected on data with a purity of $(96.98 \pm 0.01)\%$, according to MC simulations. Thus, the control sample corresponds to about 10 times the statistic expected for the signal channel.

With the requirement $M_{bc} > 5.27$ GeV applied, the distribution of C_{NB} is compared for the J/ψ inclusive MC simulated data events and data in Fig. 5.8a. An excellent agreement between MC simulation and data is seen, indicating no differences between MC simulation and data which affects the continuum suppression network. To estimate possible differences between MC simulation and data due to the $C_{NB} > 0$

requirement applied on the signal channel, the ratio of selected events on MC simulation and data is compared for different requirements on C_{NB} in Fig. 5.8b. The ratio of selected events is consistent with one, indicating no systematic difference between MC simulation and data due to this requirement. The statistical uncertainty on this comparison will be treated as a systematic uncertainty.

The control sample will be used for other cross-checks and systematic studies described in the following chapters of this thesis.

Reference

1. J. Beringer et al., (Particle Data Group). The review of particle physics. Phys. Rev. D, 86:010001, 2012. Online version available at http://pdg.lbl.gov

Chapter 6
Maximum Likelihood Fit Model

In this chapter the model used in the ML fit is described in detail. After a brief and general comment on the parametrization, the individual components of the ML are explained.

6.1 General Parametrization

Following Eq. (4.4), an unbinned extended ML fit is used to extract the 26 parameters related to polarization and *CP* violation defined in Eqs. (2.28) and (2.29), and denoted $\vec{\mu}$ in the following. All remaining parameters, such as those related to PDF shapes, are denoted by $\vec{\vartheta}$. Three components ($N_c = 3$) are included in the fit model: the signal decay $\mathrm{B}^0 \rightarrow \phi\mathrm{K}^*$ ($i = 1$), peaking background from $\mathrm{B}^0 \rightarrow \mathrm{f}_0(980)\mathrm{K}^*(892)^0$ decays ($i = 2$), and combinatorial background ($i = 3$). The corresponding yields N_i are floated parameters. Each event j is characterized by a nine-dimensional set of observables $\vec{x}_j = \{M_{\mathrm{bc}}, \Delta E, C'_{\mathrm{NB}}, M_{KK}, M_{K\pi}, \cos\theta_1, \cos\theta_2, \Phi, Q\}$, with the beam-energy-constrained mass M_{bc}, the energy difference ΔE, the transformed continuum network output C'_{NB}, the invariant mass of the ϕ candidate M_{KK}, the invariant mass of the K^* candidate $M_{K\pi}$, the three helicity angles $\cos\theta_1$, $\cos\theta_2$ and Φ, and the charge $Q = \pm 1$ of the primary kaon from the B meson, denoting the B meson flavor.

If not stated otherwise, the PDF $\mathcal{P}_i(\vec{x}_j; \vec{\mu}; \vec{\vartheta})$ for a given component i is constructed as a joint PDF of the distributions of the observables in $\vec{x}_j$. With a few exceptions, explained below, no significant linear or non-linear dependence among the fit observables has been found, using the method described in Sect. 4.3. The dependencies have been studied on MC samples and cross-checked on the control sample for the signal and peaking background components, as well as on sideband data and off-resonance data for the combinatorial background component.

M. Prim, *Polarization and CP Violation Measurements*, Springer Theses, 65
DOI: 10.1007/978-3-319-05756-9_6,

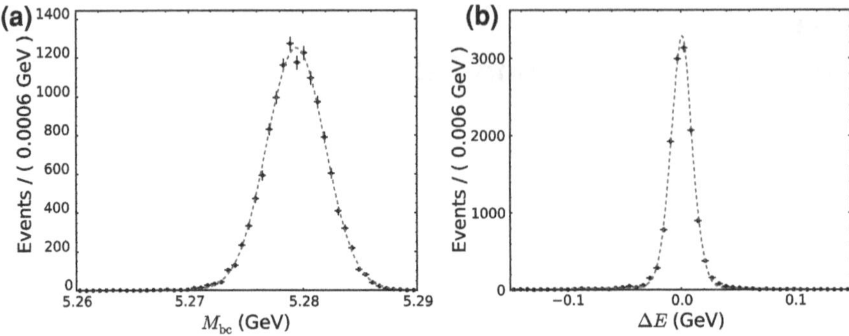

Fig. 6.1 Signal distribution of MC simulated data events (*black marker*) for **a** $M_{\rm bc}$ and **b** ΔE. The signal PDF is shown as *red dashed line*

6.2 Signal Component

The signal PDF for $\mathrm{B}^0 \rightarrow \phi \mathrm{K}^*$ is determined on MC simulated data events and cross-checked on the control sample. In the final fit, the shape parameters $\vec{\vartheta}$ are fixed to the values determined on MC simulation, except for some corrections described below.

6.2.1 $\mathbf{M_{bc}}$ *and* $\mathbf{\Delta E}$

The $M_{\rm bc}$ distribution is modeled with a double Gaussian function. The ΔE distribution is modeled with the sum of a Gaussian and two asymmetric Gaussian functions. In addition, the mean of the ΔE distribution is parametrized by a linear function of $M_{\rm bc}$, to take into account a significant linear dependence between $M_{\rm bc}$ and ΔE. The distribution on MC simulated data events and the signal PDF are shown in Fig. 6.1 for the $M_{\rm bc}$ and ΔE distribution. The linear dependence is illustrated by projections of the ΔE distribution in six bins with equal statistics of $M_{\rm bc}$ in Fig. 6.2.

The linear dependence and the conditional PDF are confirmed by ML fits to the control sample on MC simulated data events and data events. The only difference observed is due to the ΔE resolution, for which a scale factor $s = 1.124 \pm 0.062$ is derived by comparing data and MC simulated events in the control sample. The scale factor is applied to the signal model on data.

6.2.2 $\mathbf{C'_{NB}}$ *and* $\mathbf{M_{KK}}$

The $C'_{\rm NB}$ distribution is parametrized by a sum of two asymmetric Gaussian functions. The ϕ candidate mass M_{KK} is modeled by a relativistic spin-dependent BW (see Eq. (2.10)) convolved with a Gaussian function to account for resolution effects; the BW parameters can be found in Table 6.1.

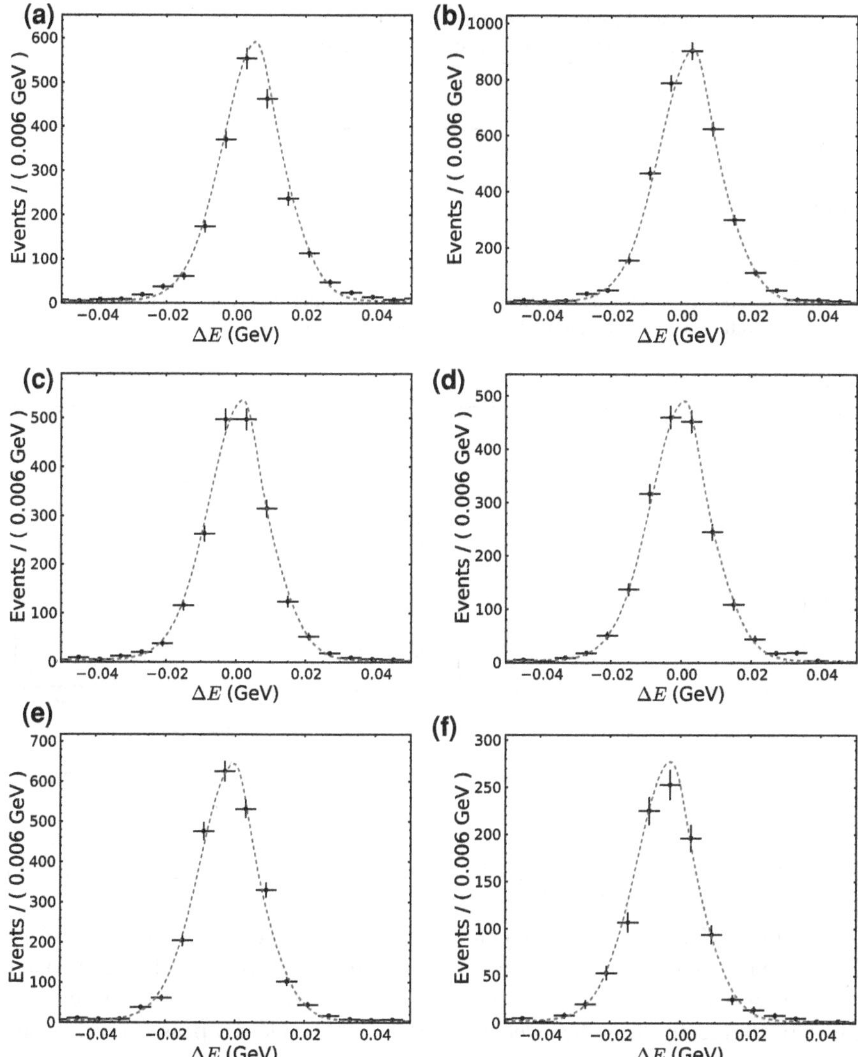

Fig. 6.2 Projection of the signal distribution of MC simulated data events (*black marker*) for ΔE in six bins of M_{bc}. The M_{bc} bin widths are given below each figure. The signal PDF is shown as *red dashed line*. From **a** to **f** the central position of the peak is shifted from *right* to *left* by about 10 MeV and illustrates the linear dependence between M_{bc} and ΔE. With respect to Fig. 6.1b, the x-axis has been limited to the interval -50 MeV $\leq \Delta E \leq$ 50 MeV to highlight the behavior. **a** 5.260 GeV $< M_{bc} \leq$ 5.277 GeV, **b** 5.277 GeV $< M_{bc} \leq$ 5.279 GeV, **c** 5.279 GeV $< M_{bc} \leq$ 5.280 GeV, **d** 5.280 GeV $< M_{bc} \leq$ 5.281 GeV, **e** 5.281 GeV < Mbc ≤ 5.283 GeV, **e** 5.283 GeV < Mbc ≤ 5.290 GeV

Table 6.1 Parameters used for the ϕ resonance are taken from Ref. [1], except for r, for which an assumption based on the values found in Kπ scattering (see Table 6.2) is made

Parameter	ϕ $J = 1$	$f_0(980)$ $J = 0$
m_J (MeV)	1019.455 ± 0.020	965 ± 10
Γ_J (MeV)	4.26 ± 0.04	...
r (GeV^{-1})	3.0 ± 1.0	...
g_π (MeV)	...	165 ± 18
g_K (MeV)	...	$(4.21 \pm 0.33)g_\pi$

Typically, values for r obtained in scattering experiments for mesons are in the order of $3\,\text{GeV}^{-1}$. For $f_0(980)$, values obtained by the BES collaboration [2] are used

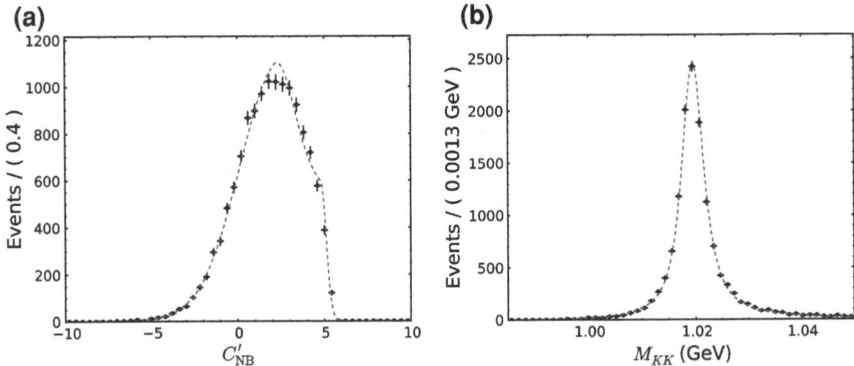

Fig. 6.3 Signal distribution of MC simulated data events (*black marker*) for **a** C'_{NB} and **b** M_{KK}. The signal PDF is shown as *red dashed line*

The distribution on MC simulated data events and the signal PDF are shown in Fig. 6.3 for the C'_{NB} and M_{KK} distribution.

As shown in Sect. 5.4, the agreement between data and MC simulation for the C'_{NB} distribution is excellent and no correction is needed. The M_{KK} resolution (about 1 MeV) differs slightly between MC simulation and data. To account for this difference, it is derived directly from the sideband data in which a clear ϕ peak is present, see also Sect. 6.4.

6.2.3 $M_{K\pi}$, $\cos\theta_1$, $\cos\theta_2$, Φ, *and* Q

The model for $M_{K\pi}$, the helicity angles, and Q is given by Eq. (2.26), which is multiplied with the experimentally derived acceptance function α $(M_{K\pi}, \cos\theta_1, \cos\theta_2, \Phi)$ to obtain the mass-angular signal PDF. The resonance parameters used for the different partial waves are given in Table 6.2.

No figures of the mass-angular distribution and PDF are shown as the distribution is not known a priori from MC simulations but has to be determined on data. However, in Fig. 6.4 the mass-angular shapes of the S-, P-, and D-wave component

Table 6.2 Resonance parameters for S-, P-, and D-wave components

Parameter	$(K\pi)_0^*$ $J=0$	$K^*(892)^0$ $J=1$	$K_2^*(1430)^0$ $J=2$
m_J (MeV)	$1435 \pm 5 \pm 5$	895.94 ± 0.22	1432.4 ± 1.3
Γ_J (MeV)	$279 \pm 6 \pm 21$	48.7 ± 0.8	109 ± 5
r (GeV^{-1})	...	3.4 ± 0.7	2.7 ± 1.3
a (GeV^{-1})	$1.95 \pm 0.09 \pm 0.06$	...	...
b (GeV^{-1})	$1.76 \pm 0.36 \pm 0.67$	...	...

The parameters m_J and Γ_J for the P- and D-wave are taken from Ref. [1], and interaction radii and S-wave parameters are taken from Ref. [3], which includes updated values with respect to Ref. [4]

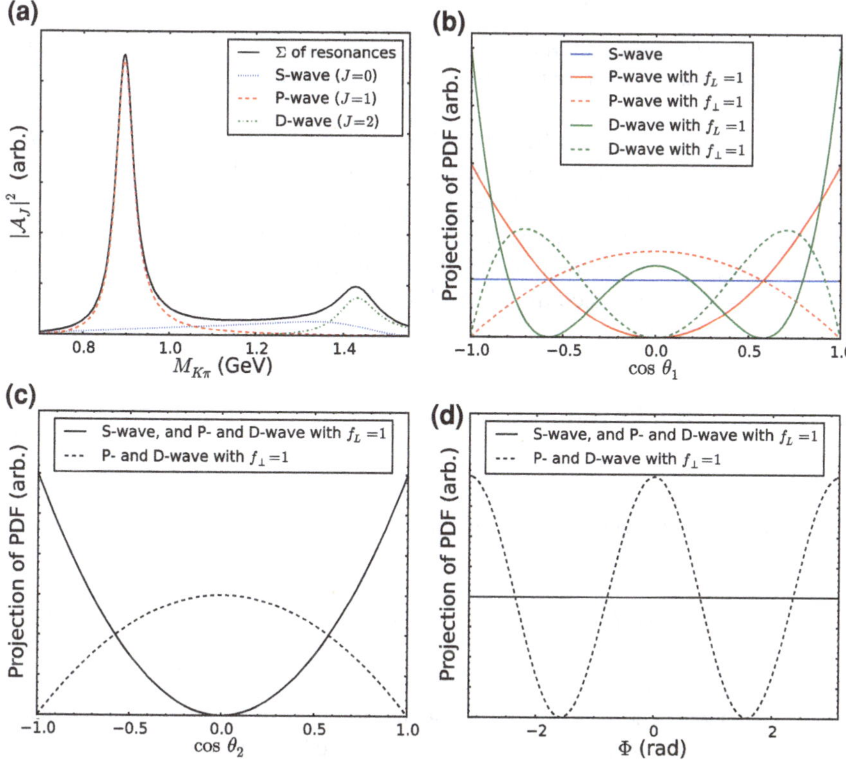

Fig. 6.4 Illustrations of **a** $M_{K\pi}$, **b** $\cos\theta_1$, **c** $\cos\theta_2$, and **d** Φ shape for the signal component. The normalization of the y-axis is arbitrary in all figures, but all angular distributions are normalized to the same value for comparison. In (**a**) the $M_{K\pi}$ lineshape of the S-, P-, and D-wave component is shown together with the incoherent sum. In (**b**) the angular distribution of $\cos\theta_1$ is illustrated for the S-, P-, and D-wave component, with two exemplary polarizations for P- and D-wave. The figure also illustrates the relation between the spin of the partial waves and the number of roots of the spheric harmonics. In (**c**) the $\cos\theta_2$ and in (**d**) the Φ shape is shown. It does not differ among partial waves but is sensitive to their polarization

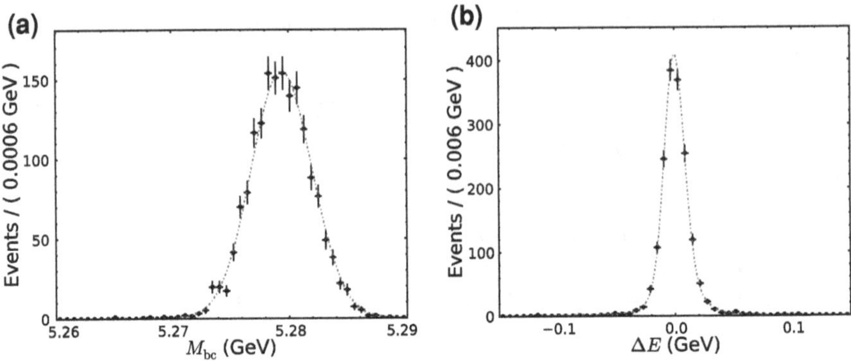

Fig. 6.5 Peaking background distribution of MC simulated data events (*black marker*) for **a** M_{bc} and **b** ΔE. The peaking background PDF is shown as *blue dash-dotted line*

are shown in projections on the observables without the correction of acceptance effects. Two exemplary polarizations, pure longitudinal polarization $f_L = 1$ and pure perpendicular polarization $f_\perp = 1$, are selected to illustrate the sensitivity of the angular observables to the polarization of the S-, P-, and D-wave component.

Neglecting possible resolution effects in the mass-angular PDF was found to have a negligible effect, according to MC simulations, and is further discussed as a systematic uncertainty.

6.3 Peaking Background Component

The peaking background PDF for $B^0 \to f_0(980)K^*(892)^0$ is also determined on MC simulated data events and all shape parameters $\vec{\vartheta}$ are fixed in the final fit.

6.3.1 M_{bc}, ΔE *and* C'_{NB}

The M_{bc}, ΔE and C'_{NB} distributions are parametrized by the same model as the signal component. The same linear dependence between M_{bc} and ΔE is present and the same scale factor for the ΔE resolution is applied on data.

The distribution on MC simulated data events and the peaking background PDF are shown in Fig. 6.5 for the M_{bc} and ΔE distribution. The C'_{NB} distribution is shown in Fig. 6.6a.

6.3.2 M_{KK}

The M_{KK} distribution of the $f_0(980)$ candidates is modeled with a Flatté function [5]. The resonance parameters are given in Table 6.1. Due to the extremely broad nature of the scalar resonance, the distribution is not convolved with a resolution function.

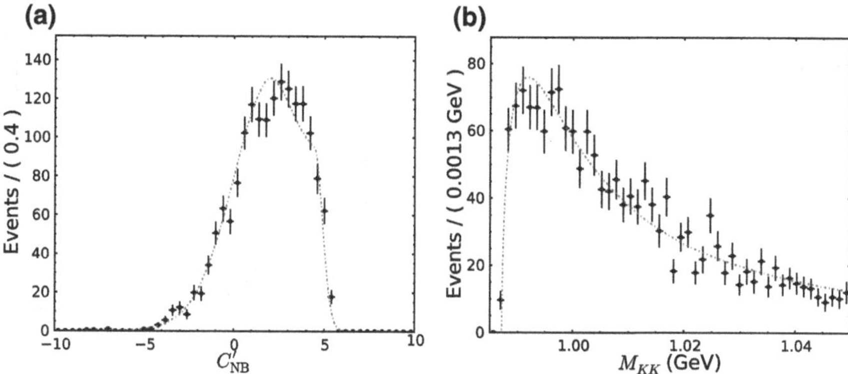

Fig. 6.6 Peaking background distribution of MC simulated data events (*black marker*) for **a** C'_{NB} and **b** M_{KK}. The peaking background PDF is shown as *blue dash-dotted line*

The available inclusive b → s MC sample does not properly simulate the coupled-channel nature of the Flatté function. Instead, a non-coupled single BW resonance, with the pole position below the K^+K^- threshold, is simulated. The two distributions are similar but the Flatté function is broader and has more events for higher values of M_{KK}.

The distribution on MC simulated data events and the peaking background PDF are shown in Fig. 6.6b for the M_{KK} distribution. For fits on MC simulated data events, the simpler BW model is used as fit function to avoid discrepancies, whereas on data the Flatté function is used. In general, the uncertainties on the nature of the broad scalar component are studied as a systematic uncertainty.

6.3.3 $M_{K\pi}$, $\cos\theta_1$, $\cos\theta_2$, Φ, *and* Q

The $M_{K\pi}$ distribution is parametrized by a relativistic spin-dependent BW for $K^*(892)^0$, using the same parameters as the signal component. The angular distribution of the peaking $B^0 \to f_0(980)K^*(892)^0$ decay, which is a pseudoscalar to scalar–vector decay, is given by Eq. 2.8. The resulting distribution is uniform in $\cos\theta_2$ and Φ, and proportional to $\cos^2\theta_1$, which is corrected for detector acceptance effects. For Q a uniform distribution is used.

The distribution on MC simulated data events and the peaking background PDF are shown in Fig. 6.7 for the $M_{K\pi}$, $\cos\theta_1$, $\cos\theta_2$, and Φ distribution.

6.4 Combinatorial Background Component

The combinatorial background PDF is determined directly from sideband data and cross-checked with the off-resonance data. Further, MC simulated data events are used to cross-check that the combinatorial background in the nominal fit region

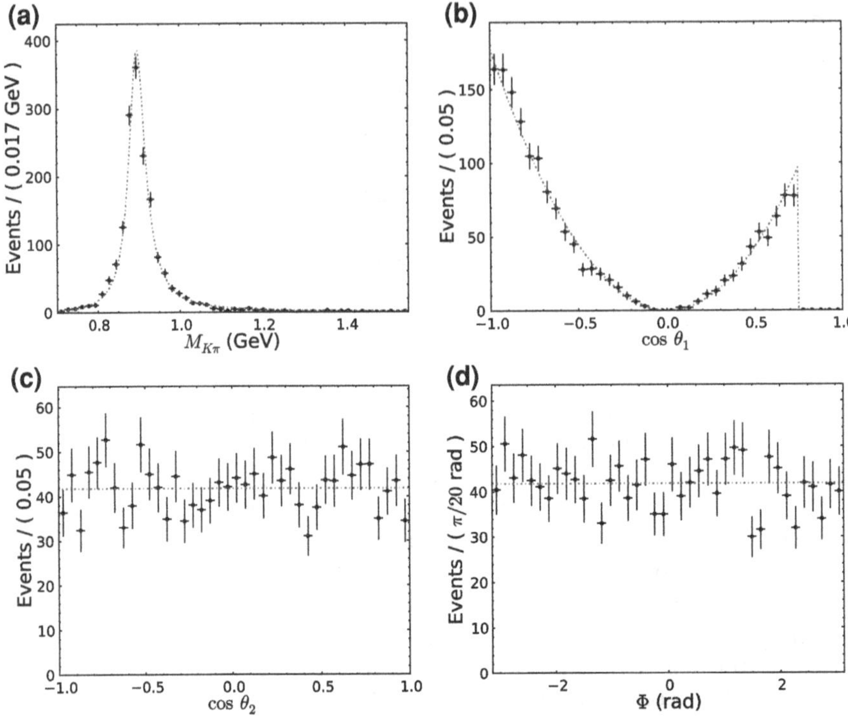

Fig. 6.7 Peaking background distribution of MC simulated data events (*black marker*) for **a** $M_{K\pi}$, **b** $\cos\theta_1$, **c** $\cos\theta_2$, and **d** Φ. The peaking background PDF is shown as *blue dash-dotted line*

is consistent with the one in the sideband region. By deriving the model directly from sideband data, the need of corrections for possible differences between MC simulation and data is avoided.

The 2 % contribution due to the combinatorial background from $\mathrm{B\overline{B}}$ events, which is present in the sideband data but missing in the off-resonance data, has no significant effect on the shape parameters and was also cross-checked with the MC simulated data samples. The shape parameters $\vec{\vartheta}$ determined on the sideband are fixed in the final fit, except for one parameter c, that is described below.

6.4.1 $\mathrm{M_{bc}}$ *and* $\Delta\mathrm{E}$

The combinatorial background PDF follows an empirically determined shape for the M_{bc} distribution, given by

$$f(M_{\mathrm{bc}}) \propto M_{\mathrm{bc}}\sqrt{1-\frac{M_{\mathrm{bc}}^2}{E_{\mathrm{beam}}^{*2}}}\exp\left[c\left(1-\frac{M_{\mathrm{bc}}^2}{E_{\mathrm{beam}}^{*2}}\right)\right], \tag{6.1}$$

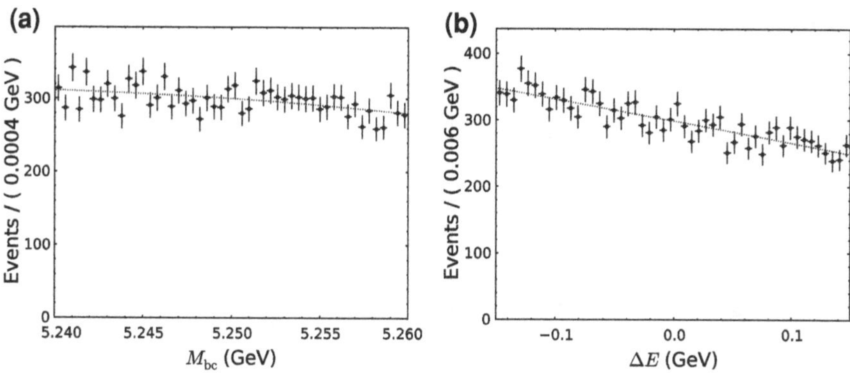

Fig. 6.8 Distribution of sideband data events (*black marker*) for **a** M_{bc} and **b** ΔE. The combinatorial background PDF is shown as *black dotted line*

where c is a free parameter. This function was first introduced by the ARGUS Collaboration [6]. The ΔE distribution is parametrized by a first-order polynomial function.

The distribution on sideband data events and the combinatorial background PDF are shown in Fig. 6.8 for the M_{bc} and ΔE distribution. The parameter c, as determined on the sideband, is not fixed in the nominal fit region but floated. Figure 5.1a illustrates the M_{bc} distribution, which differs between the sideband and the nominal fit region due to the kinematic constraints.

6.4.2 C'_{NB} *and* M_{KK}

The C'_{NB} distribution is parametrized by a sum of two asymmetric Gaussians. To account for background that contains real ϕ candidates and a non-resonant component, the M_{KK} distribution is parametrized by the sum of resonant and non-resonant contributions. Similar to the signal component, the resonant contribution is parametrized with a relativistic spin-dependent BW and convolved with the same resolution function as the signal component. The non-resonant component is described by a threshold function as

$$f(M_{KK}) \propto \arctan\left(\sqrt{(M_{KK} - 2m_K)/a}\right), \tag{6.2}$$

where m_K is the $K^{\pm}$ mass that determines the K^+K^- threshold, and a is a free parameter.

The distribution on sideband data events and the combinatorial background PDF are shown in Fig. 6.9 for the C'_{NB} and M_{KK} distribution.

Due to the presence of a clear ϕ peak in these events, the M_{KK} resolution on data events (about 1 MeV) can be determined from the sideband data and used for the

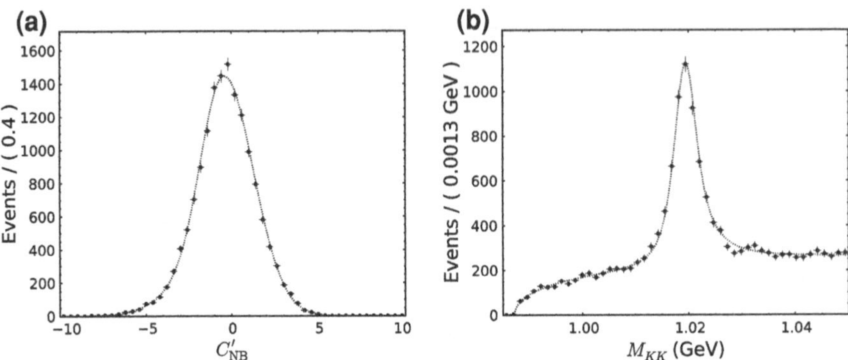

Fig. 6.9 Distribution of sideband data events (*black marker*) for **a** $C'_{\rm NB}$ and **b** M_{KK}. The combinatorial background PDF is shown as *black dotted line*

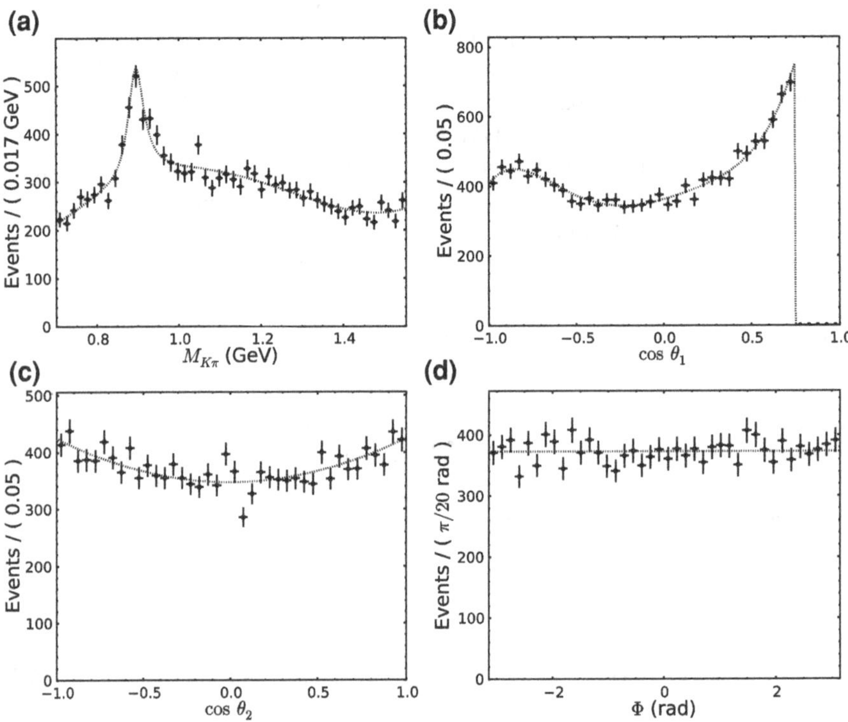

Fig. 6.10 Distribution of sideband data events (*black markers*) for **a** $M_{K\pi}$, **b** $\cos\theta_1$, **c** $\cos\theta_2$, and **d** Φ. The combinatorial background PDF is shown as *black dotted line*

signal component. This avoids possible systematic uncertainties from differences between the resolution on MC simulated data events and data events.

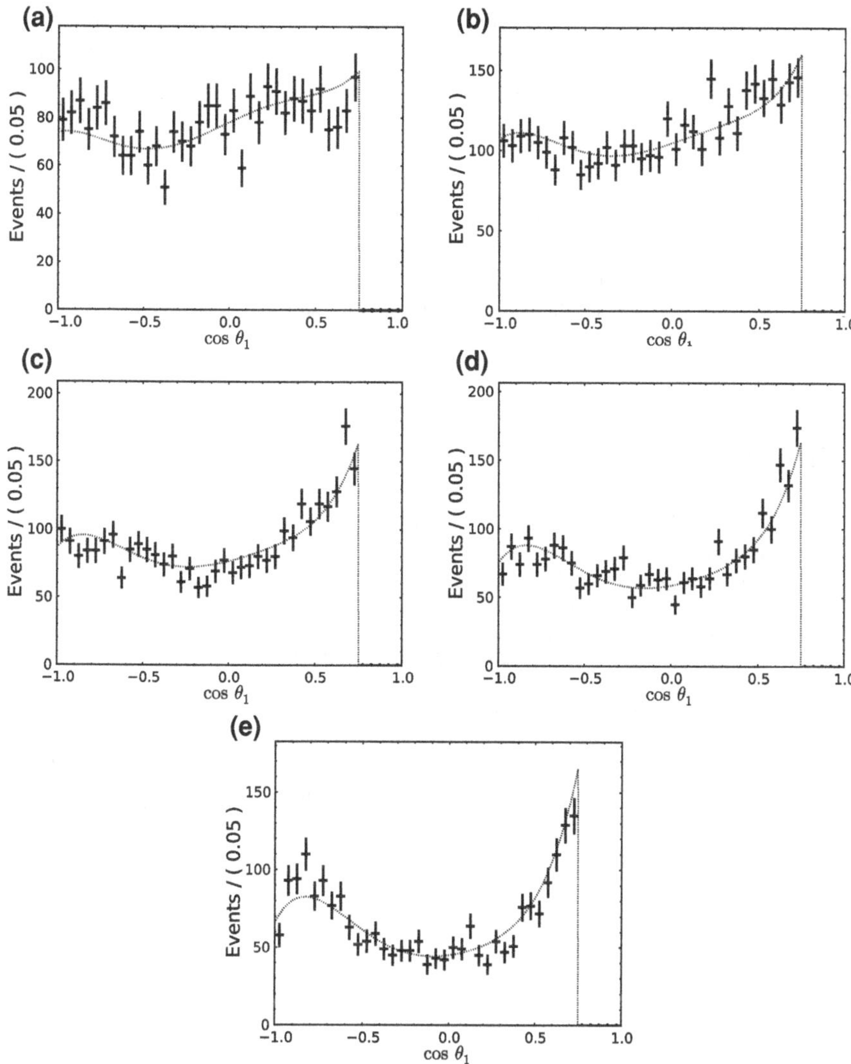

Fig. 6.11 Projection of the distribution of sideband data events (*black marker*) for $\cos\theta_1$ in five bins of $M_{K\pi}$. The $M_{K\pi}$ bin widths are given below each figure. The combinatorial background PDF is shown as *black dotted line*. **a** $0.7\,\mathrm{GeV} < M_{K\pi} \leq 0.87\,\mathrm{GeV}$, **b** $0.87\,\mathrm{GeV} < M_{K\pi} \leq 1.04\,\mathrm{GeV}$, **c** $1.04\,\mathrm{GeV} < M_{K\pi} \leq 1.21\,\mathrm{GeV}$, **d** $1.21\,\mathrm{GeV} < M_{K\pi} \leq 1.38\,\mathrm{GeV}$, **e** $1.38\,\mathrm{GeV} < M_{K\pi} \leq 1.55\,\mathrm{GeV}$

6.4.3 $\mathbf{M_{K\pi}}$, $\cos\theta_1$, $\cos\theta_2$, Φ, *and* Q

The $M_{K\pi}$ distribution is parametrized by a sum of resonant and non-resonant components. The resonant component from $K^*(892)^0$ is modeled with a relativistic spin-dependent BW using the same parameters as the signal component. The non-resonant contribution is parametrized by a fourth-order Chebyshev polynomial. A significant non-linear dependence between $M_{K\pi}$ and $\cos\theta_1$ is found in the non-resonant component of the combinatorial background. The resonant component in $M_{K\pi}$ is uniform in $\cos\theta_1$, whereas the non-resonant contribution is parametrized by a fifth-order Chebyshev polynomial, where the parameters depend linearly on $M_{K\pi}$. The $\cos\theta_2$ distribution is parametrized by a second-order Chebyshev polynomial and the distributions in Φ and Q are uniform.

The distribution on sideband data events and the combinatorial background PDF are shown in Fig. 6.10 for the $M_{K\pi}$, $\cos\theta_1$, $\cos\theta_2$ and Φ distribution. The dependence between $M_{K\pi}$ and $\cos\theta_1$ is illustrated by projections of the $\cos\theta_1$ distribution in five bins of $M_{K\pi}$ in Fig. 6.11.

References

1. J. Beringer et al., (Particle Data Group). The review of particle physics. Phys. Rev. D **86**, 010001 (2012), http://pdg.lbl.gov
2. M. Ablikim et al., (BES Collaboration). Resonances in $J/\psi \rightarrow \phi\pi^+\pi^-$ and ϕK^+K^-. Phys. Lett. B **607**(3–4), 243–253 (2005)
3. B. Aubert et al., (BaBar Collaboration). Time-dependent and time-intergrated angular analysis of $B \rightarrow \phi K_s^0\pi^0$ and $\phi K^\pm\pi^\mp$. Phys. Rev. D **78**, 092008 (2008)
4. D. Aston et al., (LASS Collaboration). A Study of $K^-\pi^+$ Scattering in the Reaction $K_p^- \rightarrow K^-\pi^+n$ at 11 GeV/c. Nucl. Phys. B **296**, 493–526 (1988)
5. S.M. Flatté, Coupled-channel analysis of the $\pi\eta$ and $K\overline{K}$ systems near $K\overline{K}$ threshold. Phys. Lett. B **63**(2), 224–227 (1976)
6. H. Albrecht et al., (ARGUS Collaboration). Search for Hadronic $b \rightarrow u$ Decays. Phys. Lett. B **241**, 278–282 (1990)

Chapter 7
Measurement of $B^0 \to \phi K^*$ Decays

This chapter covers the validation of the analysis procedure as well as the branching fraction, polarization, and *CP* violation results of the measurement of $B^0 \to \phi K^*$ decays. After the results, systematic uncertainties of the measurement are discussed in detail.

7.1 Validation

The entire analysis of $B^0 \to \phi K^*$ decays is performed as a blind analysis, i. e. the entire analysis procedure is validated and fixed before being applied to data, this process is referred to as unblinding the data. Part of this validation are the cross-checks of PDFs using the control sample and off-resonance data, as described in the last chapter. The optimization of normalization integrals was cross-checked as described in Sect. 4.2. The implementation of the angular distribution, described in Sect. 2.5.1, was compared with the independent implementation in the EvtGen [1] package to check for sign errors that could manifest e. g. as a phase shift of exactly π in the result.

Beside the checks of individual components, studies to check the robustness of the entire ML fit have been performed. These studies include ensemble tests, fits using MC simulated data events, and checks for multiple solutions.

7.1.1 *Ensemble Tests*

Ensemble tests are performed to validate a fit procedure and check for intrinsic problems. For example, if the expected statistic of a data sample is too small to constrain a certain parameter of the model or if two parameters are strongly correlated, the minimization algorithm might stop in an error state and return some arbitrary

M. Prim, *Polarization and CP Violation Measurements*, Springer Theses,
DOI: 10.1007/978-3-319-05756-9_7,

result. Sometimes, the stability of a ML fit is also limited in regions where parameters are close to or beyond physical values. Such issues can be studied in an ensemble test before unblinding the data and thus detected in advance.

In an ensemble test, the measurement is not performed on MC simulated data events but on pseudo-experiments. Instead of a computationally intensive full detector simulation, the data events are generated randomly from the PDF of the fit model. The data events are then fitted with the fit model and the obtained result is compared to the values that have been used to generate the data sample. The normalized deviation of a parameter x, also referred to as pull, and given by

$$\text{pull} = \frac{x_{\text{generated}} - x_{\text{fitted}}}{\sigma_{x,\text{fitted}}}, \tag{7.1}$$

should follow a Gaussian distribution with mean zero and width one. Typically, several hundred pseudo-experiments are performed and the obtained pull distribution is compared with the expectation.

Given 26 parameters of interest, the simulation of any possible combination of parameter values exceeds the available computational resources by far, even by using pseudo-experiments. Therefore, different sets of parameters have been selected and studied in an ensemble test. This includes sets with no assumptions about the parameters, sets with assumptions based on the previous measurements by Belle [2] and BaBar [3], sets with non-negligible *CP* violation and without, and others. All these ensemble tests did not shown any evidence for intrinsic problems of the fit model and the result of one exemplary ensemble test is presented in detail.

In the ensemble test, 500 pseudo-experiments have been performed, no *CP* violation was assumed and some arbitrary polarization was selected. The yield of signal events was chosen according to expectations based on previous measurements. In Fig. 7.1 the pull distributions of the 26 parameters defined in Eqs. (2.28) and (2.29), where the parameter of the reference amplitude a_{10} was replaced by the signal yield N_{sig}, are shown. In average, the mean and width of each distribution agrees with the expectation. Some distributions show a slight deviation, this is however expected as 26 independent parameters are regarded and thus some deviations have to be expected. A too good agreement with the expectation would indicate some intrinsic problem. It was also confirmed that with different seeds of the random number generator, these deviations randomly occur in other parameters, as it is expected from a stochastic process.

7.1.2 Fits on MC Simulated Data Events

Whereas ensemble tests check for intrinsic problems of the fit model, fits on MC simulated data events are a final check before unblinding the data. Problems due to insufficiently modeled distributions or neglected dependencies among observables can not be detected by an ensemble test.

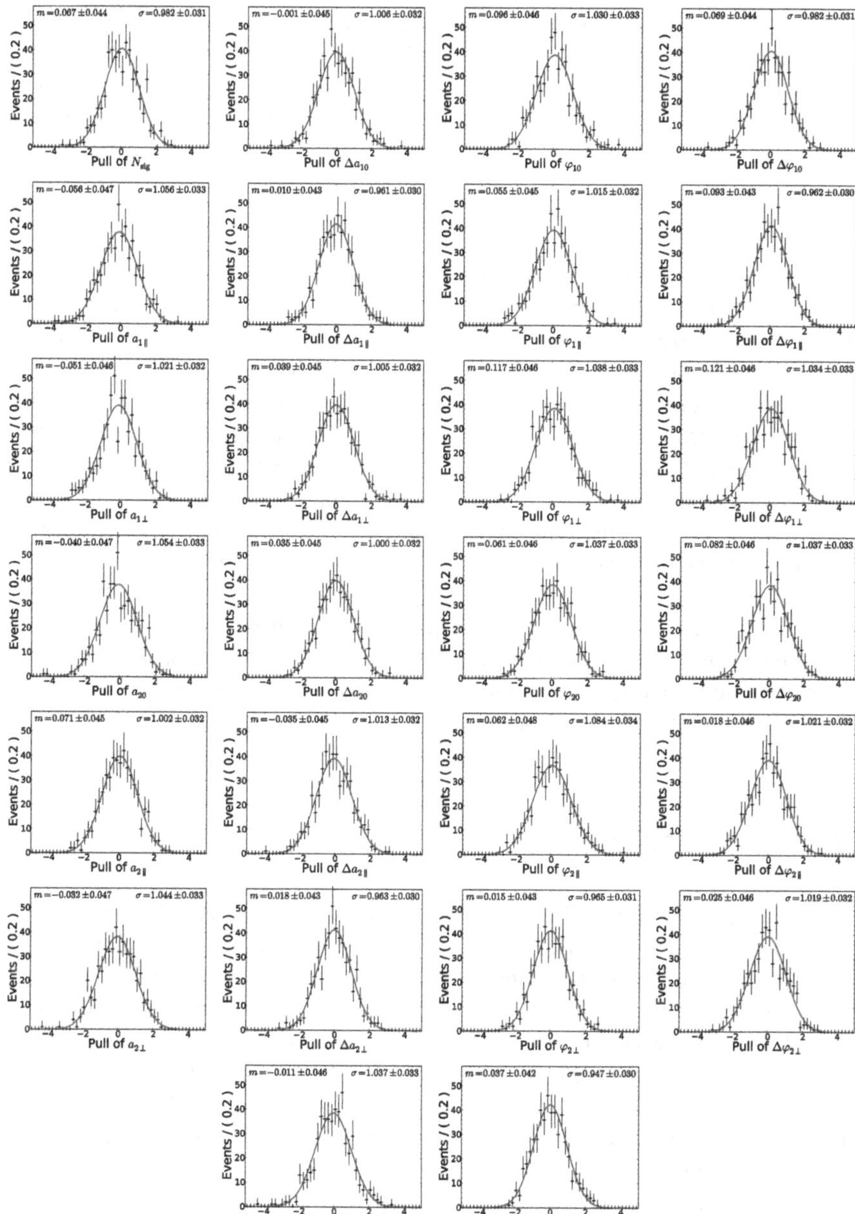

Fig. 7.1 Pull distributions (*black marker*) of the 26 parameters of interest obtained in an ensemble test. The fitted mean m and width σ of a Gaussian function (*blue solid line*) are given in the top of each figure

For the fits on MC simulated data events, the four independent streams of continuum and inclusive $b \to c$ MC, the inclusive $b \to s$ MC, and three-body $B^0 \to \phi K^+\pi^-$ phase space signal MC are used to prepare samples of MC simulated data events. The $B^0 \to \phi K^+\pi^-$ events are reweighted by assigning each event a weight

$$w = \frac{\mathrm{PDF}_{\mathrm{PHSP}}\,(M_{K\pi}, \cos\theta_1, \cos\theta_2, \Phi, Q)}{\mathrm{PDF}_{\mathrm{sig}}\,(M_{K\pi}, \cos\theta_1, \cos\theta_2, \Phi, Q)}, \tag{7.2}$$

where $\mathrm{PDF}_{\mathrm{PHSP}}$ is the PDF describing the phase space distribution and $\mathrm{PDF}_{\mathrm{sig}}$ the mass-angular signal PDF described in Sect. 6.2. By including the weights in the ML fit, arbitrary polarizations can be generated from the phase space signal MC sample, without the need of performing a full detector simulation.

The statistic is limited by the four streams of continuum and inclusive $b \to c$ MC. These samples are combined with inclusive $b \to s$ MC and reweighted signal MC events. Again, different sets of parameters, similar to the ensemble tests, are used to reweight the signal MC samples. On the prepared samples, the full analysis procedure is tested: First, the combinatorial background PDF is derived in the sideband region and afterwards the fit in the nominal fit region is used to determine the 26 parameters of interest.

In Fig. 7.2, the deviations of the 26 parameters with respect to their nominal value are shown for one exemplary set of parameters on four independent samples. Again, the fitted values agree with the expectation and indicate no problems in the fit procedure on MC simulated data events.

7.1.3 Multiple Solutions

As a last check, the possibility of multiple solutions is studied. Due to the high dimensionality of the parameter space, it is not guaranteed that the minimization algorithm finds the global minimum of the negative log-likelihood function. Pseudo-experiments are generated like in the ensemble tests and fitted with random starting values 100 times. In about 30 % of the fits, the global minimum is found. In the remaining 70 %, the algorithm is trapped in a local minimum of the negative log-likelihood function. The difference of the negative log-likelihood function at the global minimum and any local minimum fulfills $-2\Delta \ln \mathcal{L} > 50$, which excludes a local minimum by at least 7σ significance.

This behavior of the fit function is further verified with fits on MC simulated data events. Therefore, the fit on data is repeated 100 times with random starting values and the best solution is selected by the lowest negative log-likelihood value. Within 100 repetitions, it was never observed that the global minimum was not found in studies with pseudo-experiments or MC simulated data events; the lowest fraction observed has been around 25 %.

During the process of unblinding the data, the best solution was found as expected in about 30 % of all fits and all local minimums could be excluded.

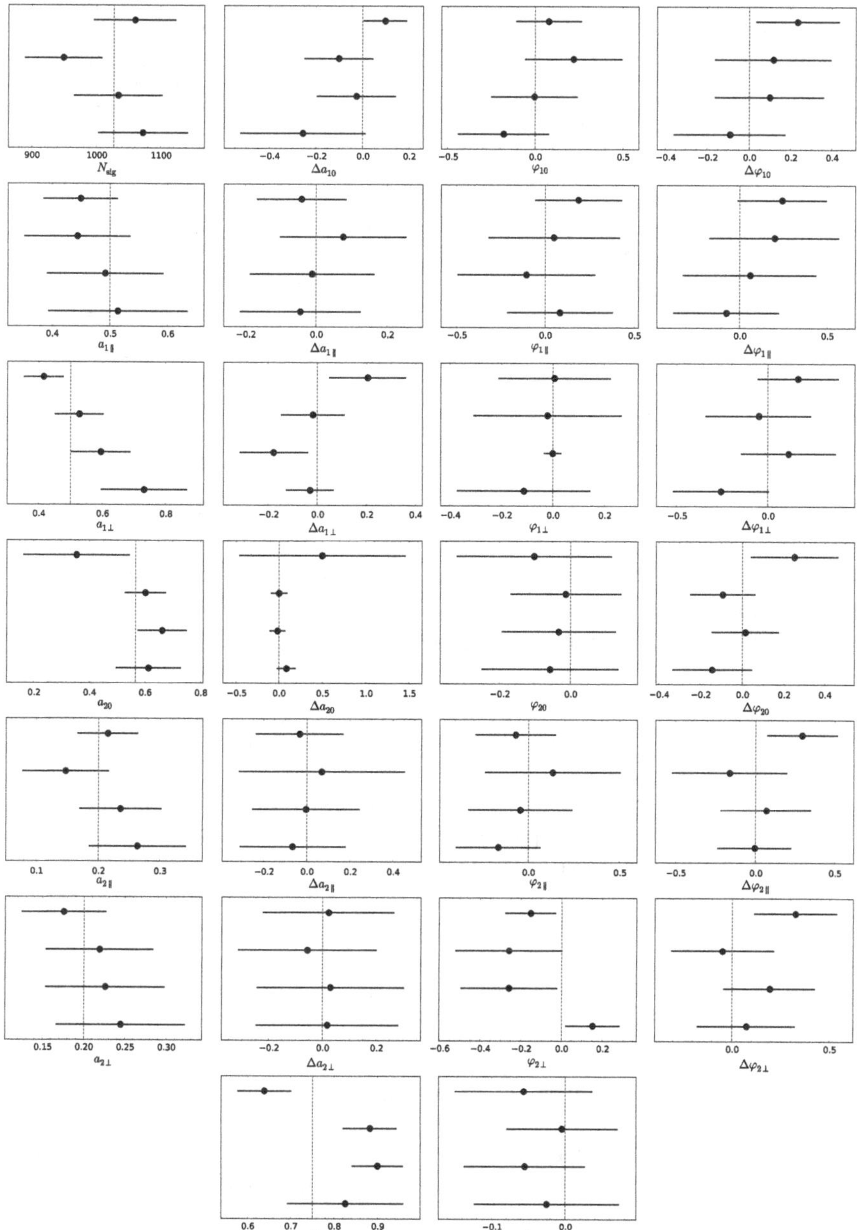

Fig. 7.2 Fitted values (*black marker*) and errors of the 26 parameters of interest obtained on four independent samples of MC simulated data events. The true value used to reweight the samples is indicated by the *blue dashed line*

Table 7.1 Summary of the branching fraction results for the $B^0 \to \phi K^*$ system

Parameter	$\phi(K\pi)^*_0$ $J = 0$	$\phi K^*(892)^0$ $J = 1$	$\phi K^*_2(1430)^0$ $J = 2$
FF_J	$0.273 \pm 0.024 \pm 0.021$	$0.600 \pm 0.020 \pm 0.015$	$0.099^{+0.016}_{-0.012} \pm 0.018$
N_J (events)	$303 \pm 29 \pm 25$	$668 \pm 34 \pm 24$	$110^{+18}_{-14} \pm 20$
$\epsilon_{reco,J}$ (%)	28.7 ± 0.1	26.0 ± 0.1	16.3 ± 0.1
ϵ_J (%)	9.4 ± 0.1	8.5 ± 0.1	2.6 ± 0.1
$\mathcal{B}_J$ (10^{-6})	$4.3 \pm 0.4 \pm 0.4$	$10.4 \pm 0.5 \pm 0.6$	$5.5^{+0.9}_{-0.7} \pm 1.0$

The branching fraction result for $B^0 \to \phi(K\pi)^*_0$ is quoted for $M_{K\pi} < 1.55$ GeV. The first error is statistical and the second due to all systematics. The error on $\epsilon_{reco,J}$ is due to MC statistics only. The error on ϵ_J is due to MC statistics and daughter branching fractions

7.2 Results

After unblinding the data sample, a signal yield of $N_{sig} = 1112 \pm 40$ events, a peaking background yield of $N_{peak} = 140 \pm 19$ events, and a combinatorial background yield of $N_{comb} = 14522 \pm 122$ events are observed, where the errors are statistical only.

The fit fraction FF_J per partial wave is defined as

$$FF_J = \frac{\int \alpha |\mathcal{A}_J|^2}{\int \alpha |\mathcal{M}|^2} = \frac{\int \alpha |\mathcal{A}_J|^2}{\int \alpha |\mathcal{A}_0 + \mathcal{A}_1 + \mathcal{A}_2|^2}, \tag{7.3}$$

where $\mathcal{A}_J$ are the partial wave amplitudes from Eqs. (2.23) to (2.25), α is the four-dimensional mass-angular acceptance function, described in Sect. 5.3, and the integral is over the full phase space. The yield per partial wave N_J is given by

$$N_J = N_{sig} FF_J \tag{7.4}$$

and the branching fraction per partial wave $\mathcal{B}_J$ by

$$\mathcal{B}_J = \frac{N_J}{\epsilon_J N_{B\overline{B}}}, \tag{7.5}$$

where ϵ_J is the product of the reconstruction efficiency $\epsilon_{reco,J}$, defined in Eq. (5.3), times the daughter branching fractions and $N_{B\overline{B}}$ is the number of $B\overline{B}$ pairs in the data sample.

The fit fraction and branching fraction results are given in Table 7.1. The sum of fit fractions is (97.2 ± 0.7) %, where the error is statistical only. This indicates the presence of a net constructive interference among the partial waves.

The results for the remaining parameters, related to polarization and *CP* violation asymmetries, are summarized in Table 7.2. The results on $B^0 \to \phi K^*(892)^0$ supersede the previous Belle result [2].

Table 7.2 Summary of the polarization and *CP* violation results for the $B^0 \to \phi K^*$ system

Parameter	$\phi(K\pi)^*_0$ $J=0$	$\phi K^*(892)^0$ $J=1$	$\phi K^*_2(1430)^0$ $J=2$
f_{LJ}	...	$0.499 \pm 0.030 \pm 0.018$	$0.918^{+0.029}_{-0.060} \pm 0.012$
$f_{\perp J}$	...	$0.238 \pm 0.026 \pm 0.008$	$0.056^{+0.050}_{-0.035} \pm 0.009$
$\phi_{\parallel J}$ (rad)	...	$2.23 \pm 0.10 \pm 0.02$	$3.76 \pm 2.88 \pm 1.32$
$\phi_{\perp J}$ (rad)	...	$2.37 \pm 0.10 \pm 0.04$	$4.45^{+0.43}_{-0.38} \pm 0.13$
δ_{0J} (rad)	...	$2.91 \pm 0.10 \pm 0.08$	$3.53 \pm 0.11 \pm 0.19$
$\mathcal{A}_{CPJ}$	$0.093 \pm 0.094 \pm 0.017$	$-0.007 \pm 0.048 \pm 0.021$	$-0.155^{+0.152}_{-0.133} \pm 0.033$
$\mathcal{A}^0_{CPJ}$	...	$-0.030 \pm 0.061 \pm 0.007$	$-0.016^{+0.066}_{-0.051} \pm 0.008$
$\mathcal{A}^\perp_{CPJ}$	...	$-0.14 \pm 0.11 \pm 0.01$	$-0.01^{+0.85}_{-0.67} \pm 0.09$
$\Delta\phi_{\parallel J}$ (rad)	...	$-0.02 \pm 0.10 \pm 0.01$	$-0.02 \pm 1.08 \pm 1.01$
$\Delta\phi_{\perp J}$ (rad)	...	$0.05 \pm 0.10 \pm 0.02$	$-0.19 \pm 0.42 \pm 0.11$
$\Delta\delta_{0J}$ (rad)	...	$0.08 \pm 0.10 \pm 0.01$	$0.06 \pm 0.11 \pm 0.02$

Parameter definitions are given in Table 2.1. The first error is statistical and the second due to systematics

Table 7.3 Triple-product correlations obtained from the weights of the $B^0 \to \phi K^*(892)^0$ partial wave, as defined in Sect. 2.5.4

	A^0_T	$A^\parallel_T$
B^0	$0.273 \pm 0.039 \pm 0.010$	$0.015 \pm 0.029 \pm 0.006$
$\overline{B}^0$	$0.210 \pm 0.039 \pm 0.014$	$0.050 \pm 0.029 \pm 0.011$
$\mathcal{A}^{0/\parallel}_T$	$0.13 \pm 0.12 \pm 0.02$	$-0.55 \pm 0.60 \pm 0.52$

The first error is statistical and the second due to systematics

All results on $B^0 \to \phi(K\pi)^*_0$, $B^0 \to \phi K^*(892)^0$, and $B^0 \to \phi K^*_2(1430)^0$ are consistent with BaBar measurements [3], with smaller errors for $B^0 \to \phi(K\pi)^*_0$ and $B^0 \to \phi K^*(892)^0$. The large longitudinal polarization fraction in the decay $B^0 \to \phi K^*_2(1430)^0$ is confirmed. Due to the requirement on $\cos\theta_1$ and the large longitudinal polarization in $B^0 \to \phi K^*_2(1430)^0$ a proportionally large drop in the efficiency with respect to the other channels is observed, which results in larger statistical uncertainties on the related parameters. In general, all parameters related to *CP* violation in the S-, P-, and D-wave components are consistent with its absence.

The results on the triple-product correlations in $B^0 \to \phi K^*(892)^0$ are summarized for B^0 and $\overline{B}^0$, together with the asymmetries, in Table 7.3. They are consistent with SM predictions of no *CP* violation.

The ambiguity in the phase parameters $\phi_{\parallel 1}$ and $\phi_{\perp 1}$ from the previous Belle measurement is resolved. In Fig. 7.3, a scan of the negative log-likelihood as a function of $\phi_{\parallel 1}$ and $\phi_{\perp 1}$ is shown, each of which shows a single solution.

The distribution of data events and the projections of the ML fit function are illustrated in Figs. 7.4, 7.5, 7.6, 7.7, 7.8, 7.9 and 7.10 for the observables M_{bc}, ΔE, C'_{NB}, M_{KK}, $M_{K\pi}$, $\cos\theta_1$, $\cos\theta_2$, and Φ in different regions. Combined figures for $B^0 \to \phi(K^+\pi^-)^*$ and $\overline{B}^0 \to \phi(K^-\pi^+)^*$, i.e. $Q = \pm 1$, are shown as no *CP*

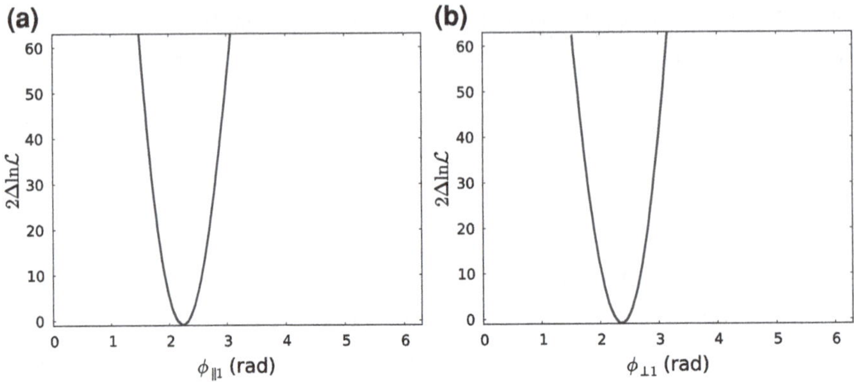

Fig. 7.3 Scan of the negative log likelihood as function of **a** $\phi_{\parallel 1}$ and **b** $\phi_{\perp 1}$. One single solution is found for each of the two phases

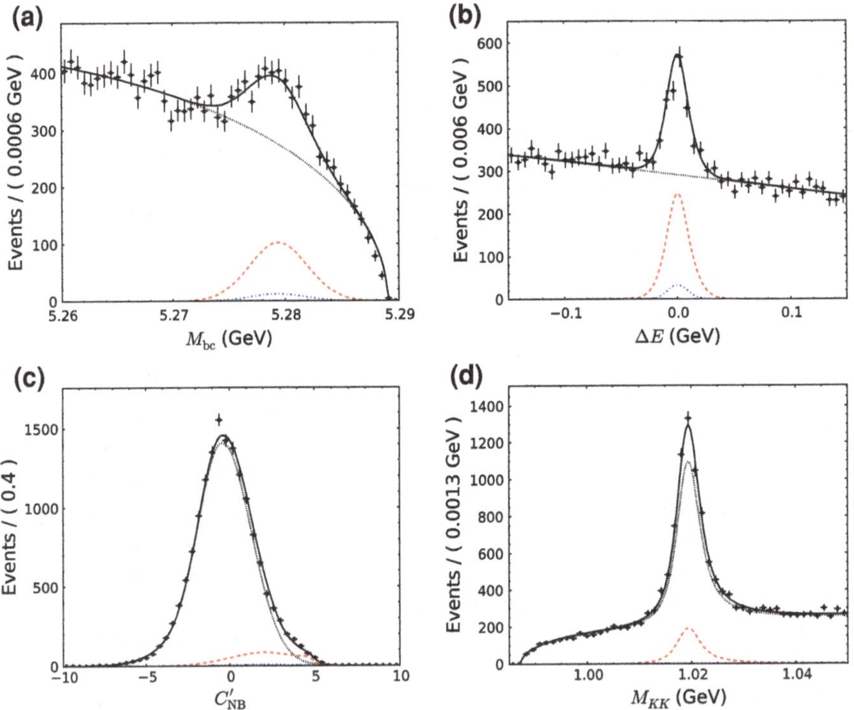

Fig. 7.4 Projections onto the observables, **a** M_{bc}, **b** ΔE, **c** C'_{NB}, and **d** M_{KK}. The data distributions are shown by *black markers*, whereas the combined ML fit function, combinatorial background, peaking background, and signal are shown by *solid black*, *dotted black*, *dash-dotted blue*, and *red curves, respectively*

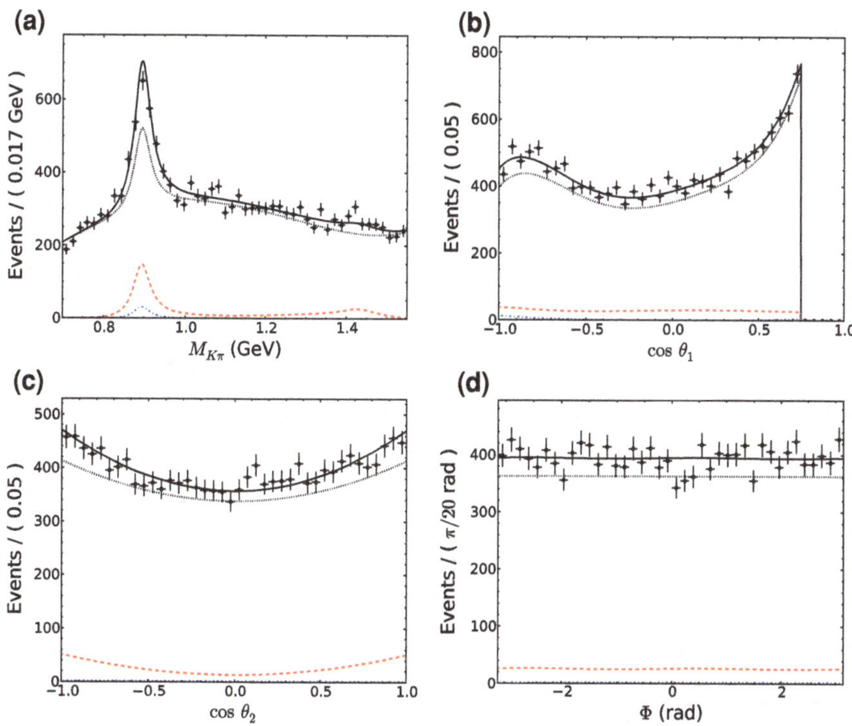

Fig. 7.5 Projections onto the observables, **a** $M_{K\pi}$, **b** $\cos\theta_1$, **c** $\cos\theta_2$, and **d** Φ. The data distributions are shown by *black markers*, whereas the combined ML fit function, combinatorial background, peaking background, and signal are shown by *solid black*, *dotted black*, *dash-dotted blue*, and *red curves*, respectively

violation is observed and independent figures would show statistically compatible distributions.

In certain figures, signal-enhancing requirements are applied; such requirements applied for each observable are $M_{\text{bc}} > 5.27\,\text{GeV}$, $-40\,\text{MeV} < \Delta E < 40\,\text{MeV}$, $1.01\,\text{GeV} < M_{KK} < 1.03\,\text{GeV}$, and $C_{\text{NB}} > -3$. Details on which requirements are applied in a specific figure are given in the caption of each figure.

In Figs. 7.4 and 7.5, the full data sample is shown by projections of each component in the ML fit. The data distributions and the combined ML fit function are in excellent agreement for all observables. In Figs. 7.6 and 7.7, the signal-enhancing requirements are applied. Again, the data distributions and combined ML fit function are in excellent agreement. In Fig. 7.8 the angular distribution of $\cos\theta_1$ is shown for three regions with enriched contributions of S-, P-, and D-wave component. In Figs. 7.9 and 7.10, the requirement $M_{\text{bc}} < 5.27\,\text{GeV}$ is applied to compare the data

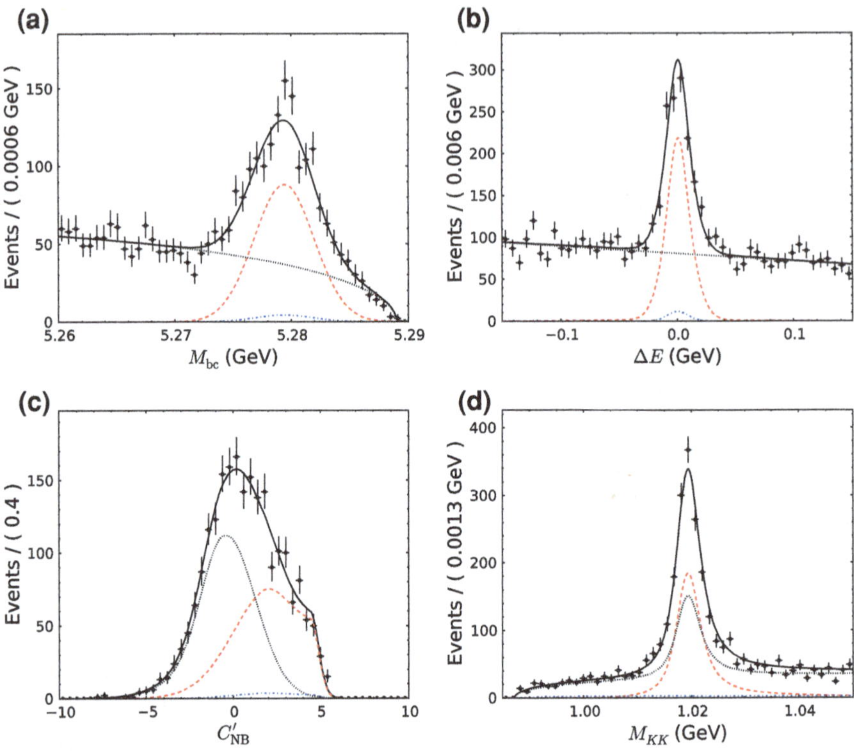

Fig. 7.6 Projections onto the observables, **a** M_{bc}, **b** ΔE, **c** C'_{NB}, and **d** M_{KK} with signal-enhancing requirements (see text) applied. In each projection, a signal-enhancing requirement on the other three observables is applied, e.g. in **a** on **b**, **c**, and **d**. The data distributions are shown by *black markers*, whereas the combined ML fit function, combinatorial background, peaking background, and signal are shown by *solid black*, *dotted black*, *dash-dotted blue*, and *red curves*, respectively

distribution with the ML fit function in a region where only combinatorial background events are expected. Also in this region, the ML fit function agrees with the distribution of data events.

7.3 Systematic Uncertainties

The systematic uncertainties on the results presented in the last section are due to various sources. In general, the systematic uncertainties can be split into two groups. The first group contains uncertainties that enter only the calculation of the branching fraction and are mainly related to uncertainties on the reconstruction efficiency. The second group covers uncertainties on the polarization and *CP* violation parameters and is rather decoupled from the first group.

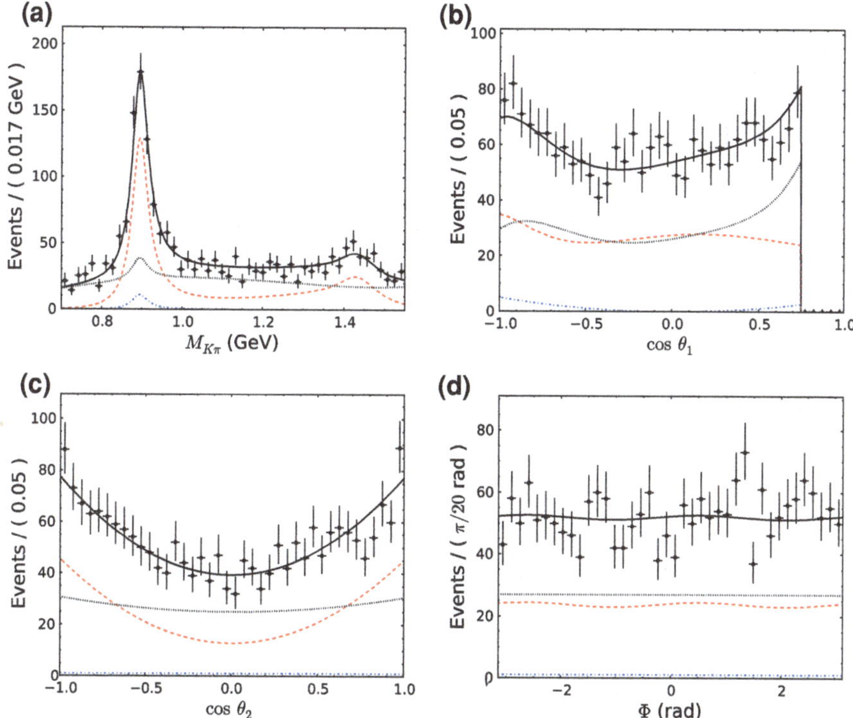

Fig. 7.7 Projections onto the observables, **a** $M_{K\pi}$, **b** $\cos\theta_1$, **c** $\cos\theta_2$, and **d** Φ with signal-enhancing requirements (see text) on $M_{\rm bc}$, ΔE, $C'_{\rm NB}$, and M_{KK} applied. The data distributions are shown by *black markers*, whereas the combined ML fit function, combinatorial background, peaking background, and signal are shown by *solid black*, *dotted black*, *dash-dotted blue*, and *red curves*, respectively

7.3.1 Branching Fraction

Track Reconstruction Efficiency

Due to uncertainties in the reconstruction efficiency of charged tracks, 0.35 % uncertainty is assigned per charged track, which results in 1.4 % total systematic uncertainty. These values have been estimated from a study, described in Ref. [4], of partially reconstructed $\mathrm{D}^{*+} \to \mathrm{D}^0\pi^+ \to (\mathrm{K}^0_{\mathrm{S}}\pi^+\pi^-)\pi^+$ decays.

PID Selection

Uncertainties on the PID requirements on kaons and pions due to MC simulation and data differences are estimated from $\mathrm{D}^{*+} \to \mathrm{D}^0\pi^+ \to (\mathrm{K}^-\pi^+)\pi^+$ samples [5]. The selection efficiencies for MC simulation and data are tabulated as a function of track momentum and polar angle. Using MC simulated signal events, a correction factor for the selection efficiency on data is determined from these tables per partial wave:

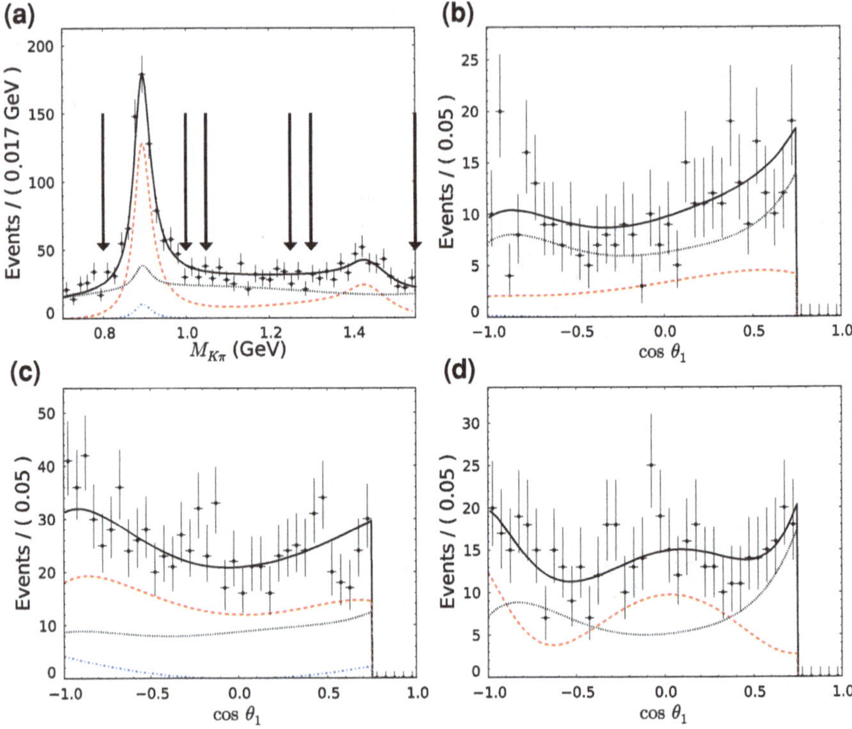

Fig. 7.8 Projections onto the observables, **a** $M_{K\pi}$ and $\cos\theta_1$ for the **b** S-, **c** P-, and **d** D-wave range with signal-enhancing requirements (see text) on M_{bc}, ΔE, C'_{NB}, and M_{KK} applied. In **b**, **c**, and **d** additional requirements on $M_{K\pi}$ are applied to enhance the contribution of the S-, P-, and D-wave contribution. The requirements are given below each figure and are illustrated by the arrows in (**a**). The data distributions are shown by *black markers*, whereas the combined ML fit function, combinatorial background, peaking background, and signal are shown by *solid black*, *dotted black*, *dash-dotted blue*, and *red curves*, respectively

$$\epsilon_{\text{corr},(K\pi)^*_0} = (98.2 \pm 3.3)\,\%,$$
$$\epsilon_{\text{corr},K^*(892)^0} = (98.2 \pm 3.3)\,\%,$$
$$\epsilon_{\text{corr},K^*_2(1430)^0} = (98.3 \pm 3.4)\,\%.$$

The correction factors are included in the reconstruction efficiencies ϵ_J given in Table 7.1. The uncertainty on the correction factor is taken as a systematic uncertainty.

C_{NB} Requirement

Possible differences in the efficiency on MC simulation and data due to the C_{NB} requirement are estimated from a comparison using the control sample. The ratio of selected events is compared for MC simulations and data events and shown in Fig. 5.8b. The uncertainty of 0.7 % on the requirement $C_{NB} > 0$ is taken as a systematic uncertainty.

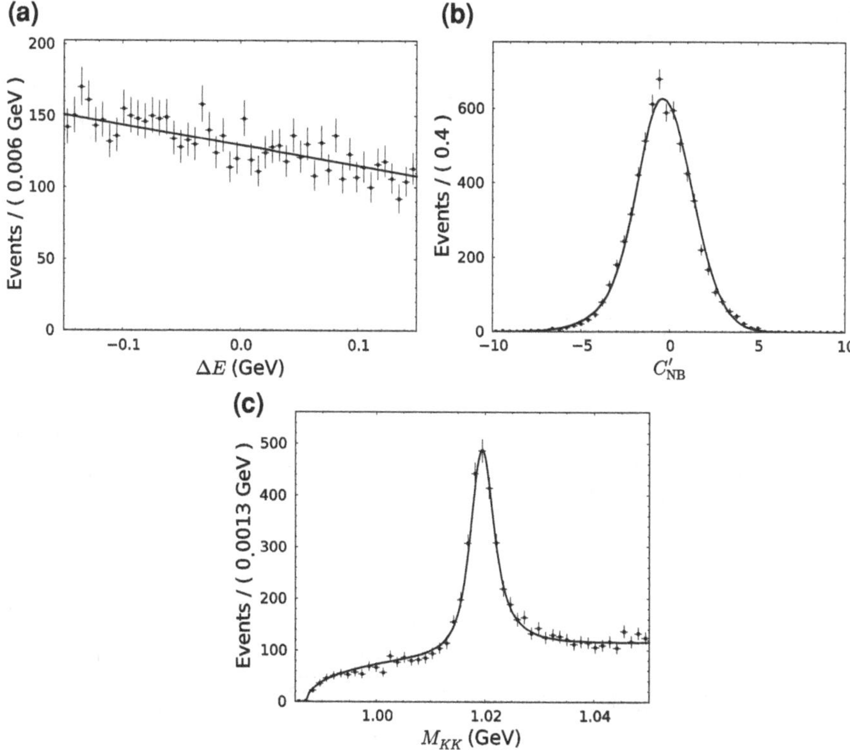

Fig. 7.9 Projections onto the observables, **a** ΔE, **b** $C'_{\rm NB}$, and **c** M_{KK} with the requirement $M_{\rm bc} < 5.27$ GeV applied. The data distributions are shown by *black markers*, whereas the curve of the combined ML fit function is congruent with the combinatorial background and shown by a *solid black line*. The *curves* of peaking background and signal vanish in this region

MC Statistics

Limited MC statistics results in a 0.5 % uncertainty on the absolute value of the reconstruction efficiency and is taken as a systematic uncertainty.

ϕ and $K_2^*(1430)$ Branching Fraction

Uncertainties on the daughter branching fractions of $\phi \to K^+K^-$ and $K_2^*(1430)^0 \to K^+\pi^-$ from Ref. [6] are taken as a systematic uncertainty.

Number of $B\overline{B}$ Pairs

The uncertainty on the number of $(772 \pm 11) \times 10^6$ $B\overline{B}$ pairs results in a 1.4 % systematic uncertainty on the branching fraction.

The individual sources of systematic uncertainties are summarized in Table 7.4 per partial wave, including the total uncertainty estimated by adding the individual errors in quadrature.

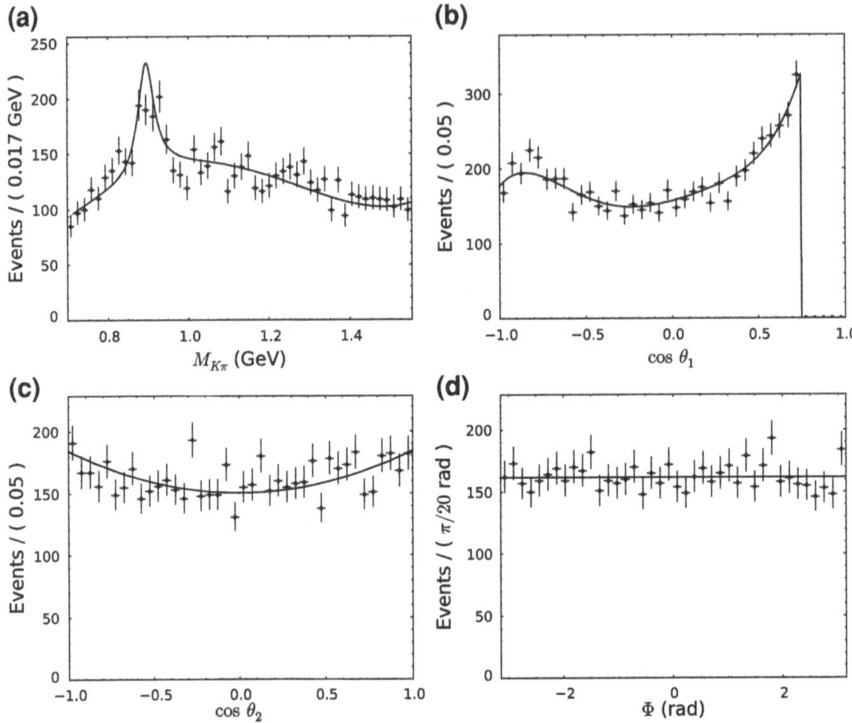

Fig. 7.10 Projections onto the observables, **a** $M_{K\pi}$, **b** $\cos\theta_1$, **c** $\cos\theta_2$, and **d** Φ with the requirement $M_{bc} < 5.27$ GeV applied. The data distributions are shown by *black markers*, whereas the *curve* of the combined ML fit function is congruent with the combinatorial background and shown by a *solid black line*. The *curves* of peaking background and signal vanish in this region

7.3.2 Polarization and CP Violation

PDF Parametrization

In the final ML fit, several parameters have been fixed. External inputs on resonance masses, widths, and other parameters are given in Tables 6.1 and 6.2 together with their uncertainties. Besides these values, shape parameters of the combinatorial background from the fit to data sideband as well as signal and peaking background shape parameters from fits to MC simulated samples and the control sample have been fixed.

All fixed parameters are varied one by one by $\pm 1\sigma$, with σ being their uncertainty, and the differences with respect to the nominal fit result are added in quadrature. The resulting error is assigned as a systematic uncertainty due to PDF parametrization. For most parameters this uncertainty is dominated by the uncertainty on the external inputs.

Table 7.4 Systematic errors (%) that enter only the calculation of the branching fraction

Source	$\phi(K\pi)^*_0$ $J = 0$	$\phi K^*(892)^0$ $J = 1$	$\phi K^*_2(1430)^0$ $J = 2$
Track reconstruction efficiency	1.4	1.4	1.4
PID selection	3.3	3.3	3.4
C_{NB} requirement	0.7	0.7	0.7
MC statistics	0.5	0.5	0.5
ϕ branching fraction	1.0	1.0	1.0
$K^*_2(1430)$ branching fraction	...	...	2.4
Number of $B\overline{B}$ pairs	1.4	1.4	1.4
Total	4.1	4.1	4.8

Resolution

The mass-angular PDF of the signal component, as well as for the other components, neglects resolution effects in $M_{K\pi}$, $\cos\theta_1$, $\cos\theta_2$, and Φ, whereas they are included in the description of M_{KK}. To estimate the systematic uncertainty from neglecting resolution effects, pseudo-experiments are generated and fitted with and without applying an additional Gaussian smearing with the resolution derived from MC simulations to the pseudo-experiment data samples.
The relative difference is found to be at least of $\mathcal{O}(10^{-4})$ for all parameters and thus negligible with respect to other systematic uncertainties. No systematic uncertainty is assigned.

Efficiency Function

Uncertainties on the efficiency function are estimated by varying the efficiency function parameters one by one by $\pm 1\sigma$. Differences between the efficiency functions for B^0 and $\overline{B}^0$ are found to be smaller than the statistical uncertainties on the efficiency function. Again, the differences to the nominal fit result are added in quadrature and taken as a systematic uncertainty.

Self-Crossfeed

The impact of the remaining SCF events, which have been neglected in the final ML fit, is estimated by pseudo-experiments that are fitted with and without adding additional SCF events from MC simulated signal samples. Signal samples with a polarization corresponding to the measured results are used to address the possibility of the SCF depending on the polarization. The amount of added SCF events is chosen according to expectations from MC simulations. The mean of the residual between fits with and without additional SCF events is found to be consistent with zero and the width of the obtained residual distribution is taken as a systematic uncertainty.

K^+K^- Shape

The M_{KK} shape of the peaking background component is modeled with a Flatté function and assumed to originate from resonant $B^0 \to f_0(980)K^*(892)^0$ decays.

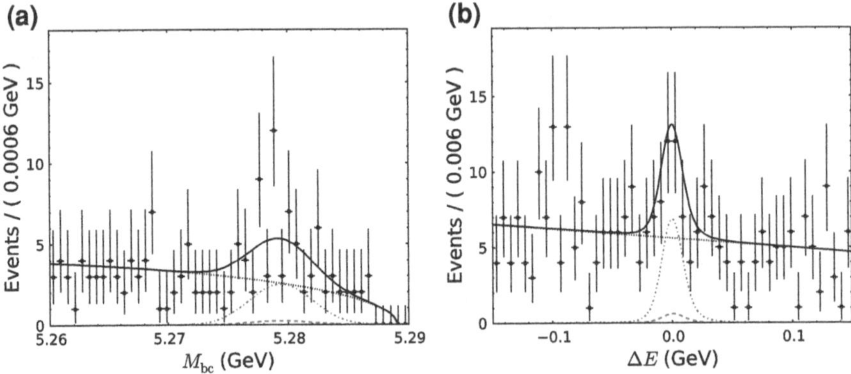

Fig. 7.11 Projections onto the observables, **a** M_{bc} and **b** ΔE for a $B^0 \to f_0(980)K^*(892)^0$ enriched region below the ϕ peak, see text for details on the applied requirements. The data distributions are shown by *black markers*, whereas the combined ML fit function, combinatorial background, peaking background, and signal are shown by *solid black*, *dotted black*, *dash-dotted blue*, and *red curves*, respectively

A possible non-resonant contribution from $B^0 \to K^+K^-K^*(892)^0$ decays is considered as a source of systematic uncertainty.

The requirements $M_{bc} > 5.27$ GeV, -40 MeV $< \Delta E < 40$ MeV, 0.985 GeV $< M_{KK} < 1.010$ GeV, $C_{NB} > -3$, and 0.8 GeV $< M_{K\pi} < 1.0$ GeV enrich potential $B^0 \to f_0(980)K^*(892)^0$ candidates from below the ϕ peak. In Fig. 7.11 the projection onto the observables M_{bc} and ΔE is shown, where the enriching requirement on the shown observable is omitted. In Fig. 7.12 the M_{KK} requirement is changed to 1.035 GeV $< M_{KK} < 1.050$ GeV to enrich the peaking background candidates from the region above the ϕ peak. Again, the enriching requirement on the shown observable is omitted. In all figures the data distribution agrees well with the fit model and does not indicate the presence of for example additional non-resonant $B^0 \to K^+K^-K^*(892)^0$ events that are expected to appear enriched in the region above the ϕ peak. In this region, the signal component has, due to the upper tail of the ϕ resonance, also a similar strong contribution as the peaking background component.

The fit model of the peaking background component is modified to allow for a coherent sum of $B^0 \to f_0(980)K^*(892)^0$ and $B^0 \to K^+K^-K^*(892)^0$ decays with relative amplitude and phase between them. Taking into account the change in the number of degrees of freedom, negative log-likelihoods obtained from this alternative fit and the nominal fit yield equally good solutions. However, the model based on the coherent sum shows a very strong destructive interference, which is also often observed in Dalitz analyses that include K^+K^- (see e.g. Ref. [7]). The nominal fit model is therefore chosen as default model and the difference with respect to the alternative model is taken as a systematic uncertainty due to general uncertainties on the nature of the scalar K^+K^- resonance. A model with only $B^0 \to K^+K^-K^*(892)^0$ is not considered as it shows a significant deviations between the data and the fit model in the M_{KK} region below the ϕ peak.

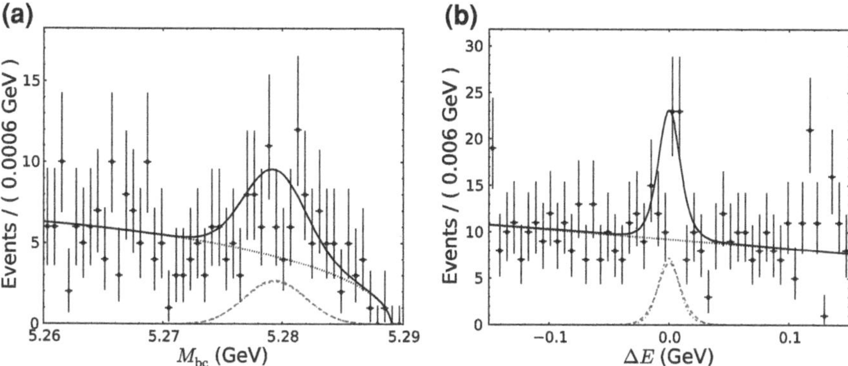

Fig. 7.12 Projections onto the observables, **a** M_{bc} and **b** ΔE for a $B^0 \to f_0(980)K^*(892)^0$ enriched region above the ϕ peak, see text for details on the applied requirements. The data distributions are shown by *black markers*, whereas the combined ML fit function, combinatorial background, peaking background, and signal are shown by *solid black*, *dotted black*, *dash-dotted blue*, and *red curves*, respectively

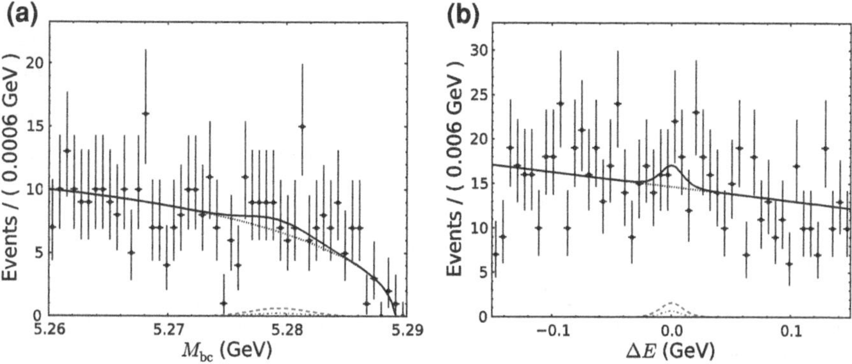

Fig. 7.13 Projections onto the observables, **a** M_{bc} and **b** ΔE for a $B^0 \to f_0(980)(K\pi)^*_0$ and $B^0 \to f_0(980)K^*_2(1430)^0$ enriched region, see text for details on the applied requirements. The data distributions are shown by *black markers*, whereas the combined ML fit function, combinatorial background, peaking background, and signal are shown by *solid black*, *dotted black*, *dash-dotted blue*, and *red curves*, respectively

Peaking Background from other K^* states

No contribution of peaking background from other K^* states is expected according to MC simulations. The requirements $M_{bc} > 5.27$ GeV, $-40\,\text{MeV} < \Delta E < 40\,\text{MeV}$, $0.985\,\text{GeV} < M_{KK} < 1.010\,\text{GeV}$, $C_{NB} > -3$, and $M_{K\pi} > 1.05$ GeV enrich the contribution of e. g. $B^0 \to f_0(980)(K\pi)^*_0$ or $B^0 \to f_0(980)K^*_2(1430)^0$ candidates. In Fig. 7.13 the projection onto the observables M_{bc} and ΔE is shown, where the enriching requirement on the shown observable is omitted. No excess of data events with respect to the nominal fit model is observed and no systematic uncertainty is assigned.

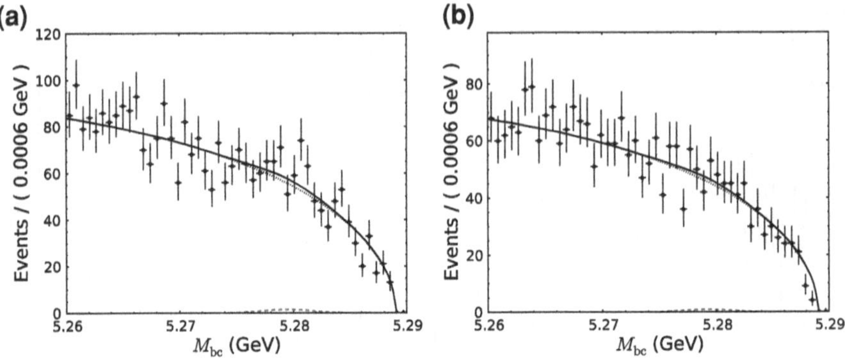

Fig. 7.14 Projections onto the observable, M_{bc} for **a** $B^0 \to \phi\phi$ and **b** $B^0 \to \phi\rho^0$ enriched regions, see text for details on the applied requirements. The data distributions are shown by *black markers*, whereas the combined ML fit function, combinatorial background, peaking background, and signal are shown by *solid black*, *dotted black*, *dash-dotted blue*, and *red curves*, respectively

Table 7.5 Systematic errors (absolute values) on the triple-product correlations for $B^0 \to \phi K^*(892)^0$, as defined in Sect. 2.5.4

Parameter	PDF	Eff.	SCF	K^+K^-	Interf.	Total
$A^0_{TB^0}$	0.003	0.003	0.003	0.002	0.008	0.010
$A^{\parallel}_{TB^0}$	0.003	0.003	0.003	0.001	0.003	0.006
$A^0_{T\overline{B}^0}$	0.004	0.001	0.006	0.000	0.012	0.014
$A^{\parallel}_{T\overline{B}^0}$	0.002	0.002	0.003	0.000	0.010	0.011
$\mathcal{A}^0_T$	0.009	0.006	0.014	0.003	0.015	0.023
$\mathcal{A}^{\parallel}_T$	0.087	0.061	0.511	0.012	0.004	0.522

The uncertainties are due to PDF parametrization, efficiency function, SCF, uncertainties on the K^+K^- shape and K^+K^- interference effects

Peaking Background from other sources

Of the peaking background sources with one track being misidentified $B^0 \to \phi\phi \to (K^+K^-)(K^+K^-)$ and $B^0 \to \phi\rho^0 \to (K^+K^-)(\pi^+\pi^-)$ are the most dominant ones, although no statistical significant contribution is expected according to MC simulations, see Sect. 5.2.

The requirements $1.01\,\text{GeV} < M_{KK} < 1.03\,\text{GeV}$, $C_{\text{NB}} > -3$, and $-150\,\text{MeV} < \Delta E < -40\,\text{MeV}$ ($40\,\text{MeV} < \Delta E < 150\,\text{MeV}$) enrich potential $B^0 \to \phi\phi$ ($B^0 \to \phi\rho^0$) candidates. In Fig. 7.14 the projection onto M_{bc} is shown for both decay modes and the data distribution is in excellent agreement with the expectation from the combinatorial background component. No systematic uncertainty is assigned.

K^+K^- Interference Effects

As the peaking background component $B^0 \to f_0(980)K^*(892)^0$ has the same final state as the analyzed $B^0 \to \phi K^*$ signal, there exists the possibility of interference

Table 7.6 Systematic errors (absolute values) on the physics parameters defined in Table 2.1

Parameter	PDF	Eff.	SCF	K^+K^-	Interf.	Charge	Total
N_{sig}	25.8	1.4	2.9	10.7	0.8	...	28.1
FF_0	0.021	0.002	0.002	0.003	0.002	...	0.021
$\mathcal{A}_{CP0}$	0.008	0.003	0.006	0.001	0.005	0.012	0.017
FF_1	0.013	0.007	0.001	0.004	0.002	...	0.015
$\mathcal{A}_{CP1}$	0.004	0.002	0.003	0.002	0.016	0.012	0.021
FF_2	0.017	0.005	0.001	0.001	0.001	...	0.018
$\mathcal{A}_{CP2}$	0.025	0.012	0.013	0.001	0.000	0.012	0.033
f_{L1}	0.005	0.016	0.002	0.005	0.002	...	0.018
$f_{\perp 1}$	0.003	0.007	0.001	0.003	0.001	...	0.008
$\phi_{\parallel 1}$	0.015	0.005	0.009	0.002	0.010	...	0.020
$\phi_{\perp 1}$	0.014	0.005	0.013	0.004	0.037	...	0.042
δ_{01}	0.078	0.018	0.006	0.007	0.011	...	0.081
$\mathcal{A}^0_{CP1}$	0.003	0.002	0.003	0.001	0.005	...	0.007
$\mathcal{A}^{\perp}_{CP1}$	0.006	0.004	0.004	0.001	0.008	...	0.011
$\Delta\phi_{\parallel 1}$	0.009	0.004	0.005	0.001	0.005	...	0.012
$\Delta\phi_{\perp 1}$	0.008	0.005	0.012	0.001	0.010	...	0.018
$\Delta\delta_{01}$	0.006	0.004	0.006	0.001	0.001	...	0.010
f_{L2}	0.011	0.006	0.003	0.000	0.000	...	0.012
$f_{\perp 2}$	0.008	0.003	0.003	0.001	0.000	...	0.009
$\phi_{\parallel 2}$	0.138	0.072	1.314	0.009	0.017	...	1.323
$\phi_{\perp 2}$	0.121	0.049	0.010	0.007	0.013	...	0.131
δ_{02}	0.177	0.053	0.010	0.002	0.002	...	0.185
$\mathcal{A}^0_{CP2}$	0.008	0.001	0.002	0.000	0.000	...	0.008
$\mathcal{A}^{\perp}_{CP2}$	0.077	0.020	0.030	0.010	0.002	...	0.085
$\Delta\phi_{\parallel 2}$	0.254	0.062	0.979	0.010	0.017	...	1.014
$\Delta\phi_{\perp 2}$	0.101	0.023	0.013	0.006	0.014	...	0.106
$\Delta\delta_{02}$	0.011	0.003	0.009	0.001	0.003	...	0.015
N_{sig} (%)	2.3	0.1	0.3	1.0	0.1	...	2.5
FF_0 (%)	7.7	0.7	0.7	1.1	0.7	...	7.9
FF_1 (%)	2.2	1.2	0.2	0.7	0.3	...	2.6
FF_2 (%)	17.2	5.1	1.0	1.0	1.0	...	18.0

The fit fraction per partial wave FF_J is defined in Eq. (7.3). In addition, the relative errors on parameters that enter the calculation of the branching fraction are shown. The uncertainties are due to PDF parametrization, efficiency function, SCF, uncertainties on the K^+K^- shape, K^+K^- interference effects and charge asymmetry in the reconstruction

effects between the S- and P-wave K^+K^- components $f_0(980)$ and ϕ, respectively. These interference effects can be treated in a similar manner to those in the $K^+\pi^-$ invariant mass by combining the signal and peaking background amplitudes with their corresponding angular distribution in the matrix element, thus leading to a full partial wave analysis of $B^0 \to (K^+K^-)(K^+\pi^-)^*$ decays.

To estimate the systematic uncertainty from neglecting these interferences, the interference term of $B^0 \to f_0(980)K^*(892)^0$ and $B^0 \to \phi K^*(892)^0$ is added to the matrix element. The interference terms of $B^0 \to f_0(980)K^*(892)^0$ with

$B^0 \to \phi(K\pi)^*_0$ and $B^0 \to \phi K^*_2(1430)^0$ are neglected as there is little overlap between the amplitudes of these channels. The difference of this modified fit model with respect to the nominal fit model is taken as a systematic uncertainty. The relative phase between the K^+K^- components of the S-wave and longitudinal amplitude of the P-wave in this modified fit model is determined to be

$$\varphi_{f_0(980)} - \varphi_{\phi,0} = (5.2 \pm 0.3)\ \text{rad}, \tag{7.6}$$

where the error is statistical only. The general uncertainties related to the description of scalar K^+K^- resonances effects the center value of this phase difference, therefore the modified fit model is not used as nominal fit model. However, this study of systematic uncertainties shows that a full partial wave analysis of $B^0 \to (K^+K^-)(K^+\pi^-)^*$ decays is in principle possible.

Charge Asymmetry

A charge bias in the reconstruction efficiency could effect the relative yield between B^0 and $\overline{B}^0$. It has been estimated for the analyses of $D^+ \to K^0_S K^+$ [8, 9] and $D^+ \to K^0_S \pi^+$ [10] decays and the procedure is described in the given References. The bias is found to be consistent with zero, where the uncertainty of 1.2 % in this estimate is assigned as a systematic uncertainty.

A summary of the systematic uncertainties is shown in Table 7.6, except for those uncertainties that have been found to be negligible and to which therefore no systematic uncertainty has been assigned. The total uncertainty per parameter is estimated by adding the individual errors in quadrature. All systematic uncertainties have also been evaluated with respect to their effect on the triple-product correlations and are summarized in Table 7.5.

References

1. D.J. Lange, The EvtGen particle decay simulation package. Nucl. Instr. Meth A **462**(1–2), 152–155 (2001)
2. K.-F. Chen et al., (Belle Collaboration). Measurement of polarization and triple-product correlations in $B \to \phi K^*$ decays. Phys. Rev. Lett. **94**, 221804 (2005)
3. B. Aubert et al., (BaBar collaboration). Time-dependent and time-intergrated angular analysis of $B \to \phi K^o_s \pi^o$ and $\phi K^\pm \pi^\pm$. Phys. Rev. D **78**, 092008 (2008)
4. P. Koppenburg, A measurement of the track finding efficiency using partially reconstructed D* decays. Belle Note #621 (2003)
5. S. Nishida, Study of kaon and pion identification using Inclusive D* samples. Belle Note #779 (2005)
6. J. Beringer et al., (Particle data group). The review of particle physics. Phys. Rev. D **86**, 010001 (2012). http://pdg.lbl.gov
7. Y. Nakahama et al., (Belle collaboration). Measurement of *CP*-violating asymmetries in $B^o \to K^+K^-K^o_S$ decays with time-dependent dalitz approach. Phys. Rev. D **82**(7), 073011 (2010)
8. B.R. Ko et al., (Belle collaboration). Evidence for *CP* violation in the decay $D^+ \to K^o_s \pi^+$. Phys. Rev. Lett. **109**(2), 021601 (2012)

9. B.R. Ko et al., (Belle collaboration). Erratum: evidence for *CP* violation in the decay $D^+ \rightarrow K_s^o\pi^+$. Phys. Rev. Lett. **109**(11), 119903(E) (2012)
10. B. R. Ko et al., (Belle collaboration). Search for *CP* violation in the decay $D^+ \rightarrow K_s^o K^+$. JHEP **02**, 98 (2013)

Chapter 8
Conclusion

In summary, this thesis presents an angular analysis of $B^0 \to \phi K^*$ decays and search for direct *CP* violation in these decays. The analysis was performed using the full Belle data sample, consisting of an integrated luminosity of $711\,\text{fb}^{-1}$ containing $(772 \pm 11) \times 10^6$ $B\overline{B}$ pairs collected at the $\Upsilon(4S)$ resonance at the KEKB asymmetric-energy e^+e^- collider.

A partial wave analysis of the $B^0 \to \phi K^*$ system with $\phi \to K^+K^-$ and $K^* \to K^+\pi^-$ was performed to distinguish among S-, P-, and D-wave contributions from $B^0 \to \phi(K\pi)^*_0$, $B^0 \to \phi K^*(892)^0$, and $B^0 \to \phi K^*_2(1430)^0$, respectively. The analysis is the first four-body final state partial wave analysis performed at the Belle experiment.

Several new and improved methods have been applied with respect to previous Belle measurements. Neural networks from the NeuroBayes package have been employed to obtain an observable that discriminates between signal and the dominating background from $e^+e^- \to q\bar{q}$ ($q \in \{u, d, s, c\}$) continuum events. The final extraction of physics parameters was achieved by a nine-dimensional maximum likelihood fit. Previous analyses at Belle utilized only up to eight observables. A tool has been developed to obtain a reliable measure of dependence among observables in multivariate data sets, which has been used in the analysis and published in Ref. [1].

Furthermore, a method that can improve the computation time of numeric integrations by orders of magnitude in partial wave and amplitude analyses in general was presented.

This analysis includes all interference effects among the different partial waves. The branching fraction $\mathcal{B}_J$, the longitudinal (perpendicular) polarization fraction f_{LJ} ($f_{\perp J}$), the relative phase of the parallel (perpendicular) amplitude $\phi_{\parallel J}$ ($\phi_{\perp J}$) to the longitudinal amplitude, and the strong phase difference between the partial waves δ_{0J} and a number of parameters related to direct *CP* violation are measured for each partial wave. In total 26 parameters are measured and summarized with their uncertainties in Table 8.1.

M. Prim, *Polarization and CP Violation Measurements*, Springer Theses,

DOI: 10.1007/978-3-319-05756-9_8,

Table 8.1 Summary of the 26 parameters measured in the $B^0 \to \phi K^*$ system

Parameter	$\phi(K\pi)_0^*$ $J=0$	$\phi K^*(892)^0$ $J=1$	$\phi K_2^*(1430)^0$ $J=2$
$\mathcal{B}_J$ (10^{-6})	$4.3 \pm 0.4 \pm 0.4$	$10.4 \pm 0.5 \pm 0.6$	$5.5^{+0.9}_{-0.7} \pm 1.0$
f_{LJ}	...	$0.499 \pm 0.030 \pm 0.018$	$0.918^{+0.029}_{-0.060} \pm 0.012$
$f_{\perp J}$	...	$0.238 \pm 0.026 \pm 0.008$	$0.056^{+0.050}_{-0.035} \pm 0.009$
$\phi_{\parallel J}$ (rad)	...	$2.23 \pm 0.10 \pm 0.02$	$3.76 \pm 2.88 \pm 1.32$
$\phi_{\perp J}$ (rad)	...	$2.37 \pm 0.10 \pm 0.04$	$4.45^{+0.43}_{-0.38} \pm 0.13$
δ_{0J} (rad)	...	$2.91 \pm 0.10 \pm 0.08$	$3.53 \pm 0.11 \pm 0.19$
$\mathcal{A}_{CPJ}$	$0.093 \pm 0.094 \pm 0.017$	$-0.007 \pm 0.048 \pm 0.021$	$-0.155^{+0.152}_{-0.133} \pm 0.033$
$\mathcal{A}^0_{CPJ}$	...	$-0.030 \pm 0.061 \pm 0.007$	$-0.016^{+0.066}_{-0.051} \pm 0.008$
$\mathcal{A}^{\perp}_{CPJ}$	...	$-0.14 \pm 0.11 \pm 0.01$	$-0.01^{+0.85}_{-0.67} \pm 0.09$
$\Delta\phi_{\parallel J}$ (rad)	...	$-0.02 \pm 0.10 \pm 0.01$	$-0.02 \pm 1.08 \pm 1.01$
$\Delta\phi_{\perp J}$ (rad)	...	$0.05 \pm 0.10 \pm 0.02$	$-0.19 \pm 0.42 \pm 0.11$
$\Delta\delta_{0J}$ (rad)	...	$0.08 \pm 0.10 \pm 0.01$	$0.06 \pm 0.11 \pm 0.02$

The first error is statistical and the second due to systematics

The obtained results supersede all previous Belle results [2] for $B^0 \to \phi K^*(892)^0$. The analysis also provides the first measurement related to the S- and D-wave components $B^0 \to \phi(K\pi)_0^*$ and $B^0 \to \phi K_2^*(1430)^0$, respectively, at Belle. The obtained results are consistent with other measurements [3] from the BaBar collaboration and improve the uncertainties on all parameters related to the S- and P-wave components. Naive expectations predict a dominant longitudinal polarization in the decay, which is confirmed for $B^0 \to \phi K_2^*(1430)^0$ decays but in conflict with the results obtained in $B^0 \to \phi K^*(892)^0$ decays. All parameters related to *CP* violation are consistent with its absence. The results of the presented measurement in $B^0 \to \phi K^*$ decays have been published in Ref. [4].

Although this analysis is the final word from Belle on $B^0 \to \phi K^*$ decays, it provides a baseline for measurements at the upcoming Super B-factory SuperKEKB and the Belle II experiment. With an integrated luminosity of 50 ab^{-1}, Belle II will be able to perform similar analyses of the $B^0 \to \phi K^*$ system with $\phi \to K^+K^-$ and $K^* \to K_S^0\pi^0$, which typically has an order of magnitude smaller experimental reconstruction efficiency than $K^* \to K^+\pi^-$. The neutral final state allows for time-dependent measurements of mixing-induced *CP* violation in $b \to (\bar{s}s)s$ decays.

Belle II will also enable a better understanding of the nature of the broad scalar K^+K^- component in the four-body $B^0 \to (K^+K^-)(K^+\pi^-)^*$ final state. With 50 ab^{-1}, the decay chains $B^0 \to f_0(980)K^*(892)^0$ and $B^0 \to f_0(980)K^*(892)^0$, where $f_0(980) \to \pi^+\pi^-$ and $a_0(980) \to \eta^0\pi^0$, can be reconstructed and used to constrain the nature of the scalar K^+K^- distribution. Furthermore, the increased statistics will allow performing analyses similar to the presented one that include higher K^+K^- and $K^+\pi^-$ invariant mass regions. Searches for decays as for example $B^0 \rightarrow f_2(1270)K^*(892)^0$ or $B^0 \rightarrow \phi K^*$ decays with F- and G-wave K^* contributions as in $B^0 \rightarrow \phi K_3^*(1780)^0$ and $B^0 \rightarrow \phi K_4^*(2045)^0$, respectively, will be possible.

References

1. M. Feindt, M. Prim, An algorithm for quantifying dependence in multivariate data sets. Nucl. Instr. Meth. A **698**, 84–89 (2013)
2. K.-F. Chen et al., (Belle collaboration), Measurement of polarization and triple-product correlations in $B \rightarrow \phi K^*$ decays. Phys. Rev. Lett. **94**, 221804 (2005)
3. B. Aubert et al., (BaBar collaboration), Time-dependent and time-intergrated angular analysis of $B \rightarrow \phi K_S^0 \pi^0$ and $\phi K^\pm \pi^\mp$. Phys. Rev. D **78**, 092008 (2008)
4. M. Prim et al., (Belle collaboration), Angular analysis of $B^0 \rightarrow \phi K^*$ decays and search for *CP* violation at Belle. Phys. Rev. D **88**(7), 072004 (2013)

Zeitfracht Medien GmbH
Ferdinand-Jühlke-Straße 7
99095 Erfurt, Deutschland
produktsicherheit@kolibri360.de